THE END OF AVERAGE

HOW WE SUCCEED
in a World That Values Sameness

TODD ROSE

HarperOne
An Imprint of HarperCollinsPublishers

HarperOne

HarperCollins books may be purchased for educational, business, or sales promotional use. For information please e-mail the Special Markets Department at SPsales@harpercollins.com.

HarperCollins website: http://www.harpercollins.com

FIRST EDITION

Designed by Terry McGrath

Library of Congress Cataloging-in-Publication Data

Names: Rose, Todd.
Title: The end of average : how we succeed in a world that values sameness / Todd Rose with Ogi Ogas.
Description: First Edition. | New York : HarperOne, 2016.
Identifiers: LCCN 2015031522| ISBN 9780062358363 (hardback) | ISBN 9780062358387 (e-book) | ISBN 9780062415806 (audio)
Subjects: LCSH: Individuality. | Conformity. | Average. | BISAC: PSYCHOLOGY / Applied Psychology. | MATHEMATICS / Probability & Statistics / General. | BUSINESS & ECONOMICS / General.
Classification: LCC BF697 .R6565 2016 | DDC 155.2—dc23 LC record available at http://lccn.loc.gov/2015031522

ISBN 978–0–06–235836–3

15 16 17 18 19   RRD(H)   10 9 8 7 6 5 4 3 2 1

For my mentor and friend, Kurt Fischer

In all affairs it's a healthy thing now and then
to hang a question mark on the things
you have long taken for granted.

BERTRAND RUSSELL, BRITISH PHILOSOPHER

CONTENTS

THE LOOK-ALIKE COMPETITION

In the late 1940s, the United States Air Force had a serious problem: its pilots could not keep control of their planes. Although this was the dawn of jet-powered aviation and the planes were faster and more complicated to fly, the problems were so frequent and involved so many different aircraft that the Air Force had an alarming, life-or-death mystery on its hands. "It was a difficult time to be flying," one retired airman told me. "You never knew if you were going to end up in the dirt." At its worst point, seventeen pilots crashed in a single day.[1]

The two government designations for these noncombat mishaps were *incidents* and *accidents,* and they ranged from unintended dives and bungled landings to aircraft-obliterating fatalities. At first, the military brass pinned the blame on the men in the cockpits, citing "pilot error" as the most common reason in crash reports. This judgment certainly seemed reasonable, since the planes themselves seldom malfunctioned. Engineers confirmed this time and again, testing the mechanics and electronics of the planes and finding no defects. Pilots, too, were baffled. The only thing they knew for sure was that their piloting skills were not the cause of the problem. If it wasn't human or mechanical error, what was it?

After multiple inquiries ended with no answers, officials turned their attention to the design of the cockpit itself. Back in 1926, when the army was designing its first-ever cockpit, engineers had measured the physical dimensions of hundreds of male pilots (the possibility of female pilots was never a serious consideration), and used this data to standardize the dimensions of the cockpit. For the next three decades, the size and shape of the seat, the distance to the pedals and stick, the height of the windshield, even the shape of the flight helmets were all built to conform to the average dimensions of a 1926 pilot.[2]

Now military engineers began to wonder if the pilots had gotten bigger since 1926. To obtain an updated assessment of pilot dimensions, the Air Force authorized the largest study of pilots that had ever been undertaken.[3] In 1950, researchers at Wright Air Force Base in Ohio measured more than 4,000 pilots on 140 dimensions of size, including thumb length, crotch height, and the distance from a pilot's eye to his ear, and then calculated the average for each of these dimensions. Everyone believed this improved calculation of the average pilot would lead to a better fitting cockpit and reduce the number of crashes—or almost everyone. One newly hired twenty-three-year-old scientist had doubts.

Lieutenant Gilbert S. Daniels was not the kind of person you would normally associate with the testosterone-drenched culture of aerial combat. He was slender and wore glasses. He liked flowers and landscaping and in high school was president of the Botanical Garden Club. When he joined the Aero Medical Laboratory at Wright Air Force Base straight out of college, he had never even been in a plane before. But it didn't matter. As a junior researcher, his job was to measure pilots' limbs with a tape measure.[4]

It was not the first time Daniels had measured the human body. The Aero Medical Laboratory hired Daniels because he had majored in physical anthropology, a field that specialized in the anatomy of humans, as an undergraduate at Harvard. During the first half of the twentieth century, this field focused heavily on trying to classify the personalities of groups of people according to their average body shapes—a practice known as "typing."[5] For example, many physical anthropologists believed a short and heavy body was indicative of a merry and fun-loving personality, while receding hairlines and fleshy lips reflected a "criminal type."[6]

Daniels was not interested in typing, however. Instead, his undergraduate thesis consisted of a rather plodding comparison of the shape of 250 male Harvard students' hands.[7] The students Daniels examined were from very similar ethnic and sociocultural backgrounds (namely, white and wealthy), but, unexpectedly, their hands were not similar at all. Even more surprising, when Daniels averaged all his data, the average hand did not resemble any individual's measurements. There was no such thing as an average hand size. "When I left Harvard, it was clear to me that if you wanted to design something for an individual human being, the average was completely useless," Daniels told me.[8]

So when the Air Force put him to work measuring pilots, Daniels harbored a private conviction about averages that rejected almost a century of military design philosophy. As he sat in the Aero Medical Laboratory measuring hands, legs, waists, and foreheads, he kept asking himself the same question in his head: *How many pilots really were average?*

He decided to find out. Using the size data he had gathered from 4,063 pilots, Daniels calculated the average of the ten physical dimensions believed to be most relevant for design, including height, chest circumference, and sleeve length. These formed the dimensions of the "average pilot," which Daniels generously defined as someone whose measurements were within the middle 30 percent of the range of values for each dimension. So, for example, even though the precise average height from the data was five foot nine, he defined the height of the "average pilot" as ranging from five seven to five eleven. Next, Daniels compared each individual pilot, one by one, to the average pilot.[9]

Before he crunched his numbers, the consensus among his fellow air force researchers was that the vast majority of pilots would be within the average range on most dimensions. After all, these pilots had already been preselected because they appeared to be average sized. (If you were, say, six foot seven, you would never have been recruited in the first place.) The scientists also expected that a sizable number of pilots would be within the average range on all ten dimensions. But even Daniels was stunned when he tabulated the actual number.

Zero.

Out of 4,063 pilots, not a single airman fit within the average range on all ten dimensions. One pilot might have a longer-than-average arm length, but a shorter-than-average leg length. Another pilot might have a big chest but small hips. Even more astonishing, Daniels discovered that if you picked out just *three* of the ten dimensions of size—say, neck circumference, thigh circumference, and wrist circumference—less than 3.5 percent of pilots would be average sized on all three dimensions. Daniels's findings were clear and incontrovertible. *There was no such thing as an average pilot.* If you've designed a cockpit to fit the average pilot, you've actually designed it to fit no one.[10]

Daniels's revelation was the kind of big idea that could have ended one era of basic assumptions about individuality and launched a new one. But even the biggest of ideas requires the correct interpretation.

We like to believe that facts speak for themselves, but they most assuredly do not. After all, Gilbert Daniels was not the first person to discover there was no such thing as an average person.

A MISGUIDED IDEAL

Seven years earlier, the *Cleveland Plain Dealer* announced on its front page a contest cosponsored with the Cleveland Health Museum and in association with the Academy of Medicine of Cleveland, the School of Medicine, and the Cleveland Board of Education. Winners of the contest would get $100, $50, and $25 war bonds, and ten additional lucky women would get $10 worth of war stamps. The contest? To submit body dimensions that most closely matched the typical woman, "Norma," as represented by a statue on display at the Cleveland Health Museum.[11]

Norma was the creation of a well-known gynecologist, Dr. Robert L. Dickinson, and his collaborator Abram Belskie, who sculpted the figure based on size data collected from fifteen thousand young adult women.[12] Dr. Dickinson was an influential figure in his day: chief of obstetrics and gynecology at the Brooklyn Hospital, president of the American Gynecological Society, and chairman of obstetrics at the American Medical Association.[13] He was also an artist—the "Rodin of obstetrics," as one colleague put it[14]—and throughout his career he used his talents to draw sketches of women, their various sizes and shapes, to study correlations of body types and behavior.[15] Like many scientists of his day, Dickinson believed the truth of something could be determined by collecting and averaging a massive amount of data. "Norma" represented such a truth. For Dickinson, the thousands of data points he had averaged revealed insight into a typical woman's physique—someone normal.

NORMA

In addition to displaying the sculpture, the Cleveland Health Museum began selling miniature reproductions of Norma, promoting her as the "Ideal Girl,"[16] launching a Norma craze. A notable physical anthropologist argued that Norma's physique was "a kind of perfection of bodily form," artists proclaimed her beauty an "excellent standard," and physical education instructors used her as a model for how young women *should* look, suggesting exercise based on a student's deviation from the ideal. A preacher even gave a sermon on her presumably normal religious beliefs. By the time the craze had peaked, Norma was featured in *TIME* magazine, in newspaper cartoons, and on an episode of a CBS documentary series, "This American Look," where her dimensions were read aloud so the audience could find out if they, too, had a normal body.[17]

On November 23, 1945, the *Plain Dealer* announced its winner, a slim brunette theater cashier named Martha Skidmore. The newspaper reported that Skidmore liked to dance, swim, and bowl—in other words, that her tastes were as pleasingly normal as her figure, which was held up as the paragon of the female form.[18]

Before the competition, the judges assumed most entrants' measurements would be pretty close to the average, and that the contest would come down to a question of millimeters. The reality turned out to be nothing of the sort. Less than 40 of the 3,864 contestants were average-size on just five of the nine dimensions and none of the contestants—not even Martha Skidmore—came close on all nine dimensions.[19] Just as Daniels's study revealed there was no such thing as an average-size pilot, the Norma Look-Alike contest demonstrated that average-size women did not exist either.

But while Daniels and the contest organizers ran up against the same revelation, they came to a markedly different conclusion about its meaning. Most doctors and scientists of the era did not interpret the contest results as evidence that Norma was a misguided ideal. Just the opposite: many concluded that American women, on the whole, were unhealthy and out of shape. One of those critics was the physician Bruno Gebhard, head of the Cleveland Health Museum, who lamented that postwar women were largely unfit to serve in the military, chiding them by insisting "the unfit are both bad producers and bad consumers." His solution was a greater emphasis on physical fitness.[20]

Daniels's interpretation was the exact opposite. "The tendency to think in terms of the 'average man' is a pitfall into which many persons blunder," Daniels wrote in 1952. "It is virtually impossible to find an average airman not because of any unique traits in this group but because of the great variability of bodily dimensions which is characteristic of all men."[21] Rather than suggesting that people should strive harder to conform to an artificial ideal of normality, Daniels's analysis led him to a counterintuitive conclusion that serves as the cornerstone of this book: *Any system designed around the average person is doomed to fail.*

Daniels published his findings in a 1952 Air Force Technical Note entitled *The "Average Man"?*[22] In it, he contended that if the military wanted to improve the performance of its soldiers, including its pilots, it needed to change the design of any environments in which those soldiers were expected to perform. The recommended change was radical: the environments needed to fit the individual rather than the average.

Amazingly—and to their credit—the air force embraced Daniels's arguments. "The old air force designs were all based on finding pilots who were similar to the average pilot," Daniels explained to me. "But once we showed them the average pilot was a useless concept, they were able to focus on fitting the cockpit to the individual pilot. That's when things started getting better."[23]

By discarding the average as their reference standard, the air force initiated a quantum leap in its design philosophy, centered on a new guiding principle: *individual fit.* Rather than fitting the individual to the system, the military began fitting the system to the individual. In short order, the air force demanded that all cockpits needed to fit pilots whose measurements fell within the 5 percent to 95 percent range on each dimension.[24]

When airplane manufacturers first heard this new mandate, they balked, insisting it would be too expensive and take years to solve the relevant engineering problems. But the military refused to budge, and then—to everyone's surprise—aeronautical engineers rather quickly came up with solutions that were both cheap and easy to implement. They designed adjustable seats, technology now standard in all automobiles. They created adjustable foot pedals. They developed adjustable helmet straps and flight suits. Once these and other design solutions were put into place, pilot performance soared, and the U.S. Air Force became the most dominant air force on the planet. Soon, every branch of the American military published guides decreeing that equipment should fit a wide range of body sizes, instead of standardized around the average.[25]

Why was the military willing to make such a radical change so quickly? Because changing the system was not an intellectual exercise—it was a practical solution to an urgent problem. When pilots flying faster than the speed of sound were required to perform tough maneuvers using a complex array of controls, they couldn't afford to have a gauge just out of view or a switch barely out of reach. In a setting where split-second decisions meant the difference between life and death, pilots were forced to perform in an environment that was already stacked against them.

THE HIDDEN TYRANNY OF THE AVERAGE

Imagine the good that would have resulted if, at the same time the military changed the way it thought about soldiers, the rest of our society had followed suit. Rather than comparing people to a misguided ideal, they could have seen them—and valued them—for what they are: *individuals*. Instead, today most schools, work-places, and scientific institutions continue to believe in the reality of Norma. They design their institutions and conduct their research around an arbitrary standard—the average—compelling us to com-pare ourselves and others to a phony ideal.

From the cradle to the grave, you are measured against the ever-present yardstick of the average, judged according to how closely you approximate it or how far you are able to exceed it. In school, you are graded and ranked by comparing your performance to the average student. To get admitted to college, your grades and test scores are compared to the average applicant. To get hired by an employer, your grades and test scores—as well as your skills, years of experience, and even your personality score—are compared to the average applicant. If you do get hired, your annual review will quite likely compare you, yet again, against the average employee in your job level. Even your financial opportunities are determined by a credit score that is evaluated by—you guessed it—its deviation from the average.

Most of us know intuitively that a score on a personality test, a rank on a standardized assessment, a grade point average, or a rat-ing on a performance review doesn't reflect your, or your child's, or your students', or your employees' abilities. Yet the concept of aver-age as a yardstick for measuring individuals has been so thoroughly ingrained in our minds that we rarely question it seriously. Despite our occasional discomfort with the average, we accept that it repre-sents some kind of objective reality about people.

What if I were to tell you that this form of measurement—the average—was *almost always* wrong? That when it comes to under-standing individuals, the average is most likely to give incorrect and misleading results? What if, like the cockpit designs and Norma statues, this ideal is just a myth?

The central premise of this book is deceptively simple: *no one is average*. Not you. Not your kids. Not your coworkers, or your students, or your spouse. This isn't empty encouragement or hollow sloganeering. This is a scientific fact with enormous practical consequences that you cannot afford to ignore. You might be thinking I am touting a world that sounds suspiciously like Lake Wobegon from Garrison Keillor's *Prairie Home Companion,* a place where "All the children are above average." Some people *must* be average, you might insist, as a simple statistical truism. This book will show you how even this seemingly self-evident assumption is deeply flawed and must be abandoned.

It is not that the average is never useful. Averages have their place. If you're comparing two different *groups* of people, like comparing the performance of Chilean pilots with French pilots—as opposed to comparing two *individuals* from each of those groups—then the average can be useful. But the moment you need *a* pilot, or *a* plumber, or *a* doctor, the moment you need to teach *this* child or decide whether to hire *that* employee—the moment you need to make a decision about any individual—the average is useless. Worse than useless, in fact, because it creates the illusion of knowledge, when in fact the average disguises what is most important about an individual.

In this book, you will learn that just as there is no such thing as average body size, there is no such thing as average talent, average intelligence, or average character. Nor are there average students or average employees—or average brains, for that matter. Every one of these familiar notions is a figment of a misguided scientific imagination. Our modern conception of the average person is not a mathematical truth but a human invention, created a century and a half ago by two European scientists to solve the social problems of their era. Their notion of the "Average Man" did indeed solve many of their challenges and even facilitated and shaped the Industrial

Age—but we no longer live in the Industrial Age. Today we face very different problems—and we possess science and math far better than what was available in the nineteenth century.

Over the past decade, I have been part of an exciting new interdisciplinary field of science known as the *science of the individual*.[26] The field rejects the average as a primary tool for understanding individuals, arguing instead that we can only understand individuals by focusing on individuality in its own right. Cellular biologists, oncologists, geneticists, neuroscientists, and psychologists have recently begun to adopt the principles of this new science to fundamentally transform the study of cells, disease, genes, brains, and behavior. Several of the most successful businesses have begun to implement these principles, too. In fact, the principles of individuality are starting to be applied just about everywhere except for the one place where they will have their greatest impact—in your own life.

I wrote *The End of Average* to change that.

In the chapters that follow, I'll share with you three principles of individuality—*the jaggedness principle, the context principle,* and *the pathways principle.* These principles, drawn from the latest science in my field, will help you understand what is truly unique about you and, more importantly, show you how to take full advantage of your individuality to gain an edge in life. You no longer need to fly a World War II aircraft in an age of jet fighters, and you no longer need to weigh yourself against a non-existent Norma.

THE PROMISE OF INDIVIDUALITY

We are on the brink of a new way of seeing the world, a change driven by one big idea: individuality matters. You might think it is overly simplistic to believe that such a basic notion could produce profound practical consequences. But just consider what happened when another big idea was introduced to the world: the idea of germs.

In the nineteenth century, the most respected health and medical experts all insisted that diseases were caused by "miasma," a fancy term for bad air.[27] Western society's system of heath was based on this assumption: to prevent diseases, windows were kept open or closed, depending on whether there was more miasma inside or outside the room; it was believed that doctors could not pass along disease, because gentlemen did not inhabit quarters with bad air. Then the idea of germs came along.[28]

One day, everyone believed that bad air makes you sick. Then, almost overnight, people started realizing there were invisible things called microbes and bacteria that were the real cause of diseases. This new view of disease brought sweeping changes to medicine, as surgeons adopted antiseptics and scientists invented vaccines and antibiotics. But, just as momentously, the idea of germs gave ordinary people the power to influence their own lives. Now, if you wanted to stay healthy, you could do things like wash your hands, boil your water, cook your food thoroughly, and clean cuts and scrapes with iodine.

That shift in thinking about the world is similar to how I want you to think about the old world of averages and the new world of individuality. Today we have the ability to understand individuals and their talents on a level that was not possible before. This new idea will have a profound impact on our institutions—instead of viewing talent as a scarce commodity, schools will be able to nurture excellence in every student, and employers will be able to hire and retain a wider range of high-impact employees. People who feel they have unrecognized and untapped potential, who are not getting the chance to show what they are truly capable of, will be able to live up to their own unmet expectations.

Perhaps your child has been labeled a poor reader, but rather than simply being diagnosed, his school realizes that he is following an alternative, equally valid pathway to reading and adjusts your child's instruction accordingly. Perhaps one of your employees whose performance is suffering has been labeled as "difficult to work with" by her colleagues; but rather than fire her, you are able to identify the contexts that make her act out, helping her strengthen her relationships and drastically improve her performance, and allowing you to discover a hidden gem in your department. Once you see the profound changes that can take place when you apply the principles of individuality, you won't be able to see averages in the same way again.

It's unacceptable that in an age when we can map the human genome and tweak genetic coding to improve our health, we haven't been able to accurately map human potential. My work—and the message in this book—is geared toward helping us fix that. Human potential is nowhere near as limited as the systems we have put in place assume. We just need the tools to understand each person as an individual, not as a data point on a bell curve.

I know this firsthand.

I first became interested in the idea of individuality because I was crashing over and over again in my own life, and I couldn't figure out why. No matter what I tried, it seemed like everything ended up in failure. When I was eighteen, I dropped out of high school with a 0.9 GPA—that's a D-minus average. Before I was old enough to drink, I had held ten different minimum-wage jobs while trying to support a wife and son. Another son arrived when I was twenty-one. At the lowest point in my life, I was on welfare and working as an in-home nursing assistant performing enemas for $6.45 an hour.

Almost everyone said the problem was me, that I was lazy, stupid, or—most frequently—a "troublemaker." More than one school official told my parents that they would have to temper their expectations about what I would be able to achieve in life. But even during my lowest moments, I always felt that something wasn't right with this analysis. I felt sure I had something to offer; it just seemed like there was a profound mismatch between who I really was and the way the world saw me.

At first, I felt like the solution was to strive to be the same as everyone else—but that usually ended up in disaster. I failed class after class, and departed job after job. Eventually, I decided to stop trying to conform to the system and instead focused on figuring out how to make the system fit to me. It worked: fifteen years after I dropped out of high school, I was on the faculty at the Harvard Graduate School of Education, where I am now the director of the Mind, Brain, and Education program.

My own success did not happen because I awakened some secret talent overlooked by the world. It was not simply because one day I buckled down and started working hard, or because I discovered some kind of abstract new philosophy. I didn't have time for abstract: I needed to get off welfare, provide for my children, and find a pragmatic path to a rewarding career. No, I was able to change the course of my life because I followed the principles of individuality, intuitively at first, and then with conscious determination.

I wrote *The End of Average* to share these principles with you and show you how they can help you improve your performance in school, work, and your personal life. The hardest part of learning something new is not embracing new ideas, but letting go of old ones. The goal of this book is to liberate you, once and for all, from the tyranny of the average.

PART I

THE AGE OF AVERAGE

Individual talent is too sporadic and unpredictable to be allowed any important part in the organization of society. Social systems which endure are built on the average person who can be trained to occupy any position adequately if not brilliantly.

—STUART CHASE, *THE PROPER STUDY OF MANKIND*

CHAPTER 1

THE INVENTION OF THE AVERAGE

In 2002, UC Santa Barbara neuroscientist Michael Miller conducted a study of verbal memory. One by one, sixteen participants lay down in an fMRI brain scanner and were shown a set of words. After a rest period, a second series of words was presented and they pressed a button whenever they recognized a word from the first series. As each participant decided whether he had seen a particular word before, the machine scanned his brain and created a digital "map" of his brain's activity. When Miller finished his experiment, he reported his findings the same way every neuroscientist does: by averaging together all the individual brain maps from his subjects to create a map of the Average Brain.[1] Miller's expectation was that this average map would reveal the neural circuits involved in verbal memory in the typical human brain.

Whenever you read about some new neuroscience discovery accompanied by a blob-splotched cross section of a brain—here are the regions that light up when you feel love; here are the regions that light up when you feel fear—it's a near certainty that you are looking at a map of an Average Brain. As a graduate student, I was also taught the method of producing and analyzing the Average Brain (referred to as the "random effects model" in the jargon of science[2]) when I was trained in brain imaging at Massachusetts General Hospital. The driving assumption of this method is that the Average Brain represents the normal, typical brain, while each individual brain represents a variant of this normal brain—an assumption that mirrors the one that motivated the Norma look-alike contest. This premise leads neuroscientists to reject left-handed people from their studies (since it is presumed the brains of left-handed people are different from normal brains) or sometimes even throw out those individuals whose brain activity deviates *too* far from average, since researchers worry these outliers might cloud their view of the Average Brain.

There would have been nothing strange about Miller reporting the findings of his study by publishing his map of the Average Brain. What *was* strange was the fact that when Miller sat down to analyze his results, something made him decide to look more carefully at the individual maps of his research participants' brains. Even though Miller was investigating a well-studied mental task using the standard method of brain research—and even though there was nothing unusual about his participants' Average Brain—he glanced over a few of the individual maps. "It was pretty startling," Miller told me. "Maybe if you scrunched up your eyes real tight, a couple of the individual maps looked like the average map. But most didn't look like the average map at all."[3]

Other people before Miller had noticed that individual brains often failed to resemble the Average Brain, but since everyone else ignored this awkward fact, they usually ignored it, too—just as scientists and physicians long ignored the fact that no real woman looked like Norma. But now Miller did something that might seem perfectly obvious to do, yet few had ever bothered to attempt: he systematically compared each of the sixteen individual brain maps from his verbal memory experiment to the map of the Average Brain. What he found astonished him. Not only was each person's brain different from the average, they were all different from one another.

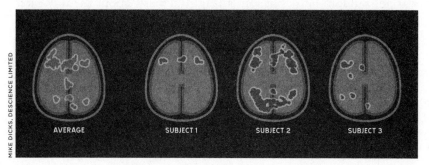

MEMORY ACTIVITY IN THE BRAIN

Some people's brains were mostly activated on the left, others on the right. Some brains were mostly activated in the front, others in the back. Some looked like a map of Indonesia with long, thick archipelagos of activation; others were almost entirely blank. But there was no avoiding the most salient fact: nobody's brain looked like the Average Brain. Miller's results paralleled those obtained by Gilbert Daniels during his investigation of hands, except this time the organ being studied wasn't a limb—it was the very cradle of thought, feeling, and personality.

Miller was bewildered. According to the key assumption behind the method of the Average Brain, most people's brains should be fairly close to average. Neuroscientists certainly expected that *some* brains should be similar to the average. But hardly any of the brains in Miller's study even remotely resembled the Average Brain. Miller feared that perhaps there had been some kind of technical error in his equipment, so he brought many of the same participants back a couple months later and once again scanned their brains as they performed the same word memory task. The results were nearly identical: each person's new brain map was pretty close to his original brain map—and each individual brain map remained quite different from the map of the Average Brain.

"That convinced me that the individual patterns we were seeing was not random noise but something systematic about the way each individual performed the task, that each person's memory system consisted of a unique neural pattern," Miller explained to me. "But what was most surprising was that these differences in patterns were not subtle, they were *extensive*."[4]

The "extensive" differences that Miller found in people's brains aren't limited to verbal memory. They've also been found in studies of everything from face perception and mental imagery to procedural learning and emotion.[5] The implications are hard to ignore: if you build a theory about thought, perception, or personality based on the Average Brain, then you have likely built a theory that applies to no one. The guiding assumption of decades of neuroscience research is unfounded. *There is no such thing as an Average Brain.*

When Miller published his counterintuitive findings, at first they elicited skepticism. Some scientists suggested the findings might be due to problems with his software algorithms, or simply bad luck in his choice of subjects—maybe too many of his participants were "outliers." The most common response from Miller's colleagues, however, was not criticism, but bored dismissal. "Other people had noticed what I noticed before in their own work; they just shrugged it off," Miller told me. "People were saying, 'Everybody already knows this, it's no big deal. That's why you use the average, it takes all these individual differences into account. You don't need to bother pointing out all this variability, because it doesn't matter.'"[6]

But Miller was convinced it *did* matter. He knew this was not some academic debate, but a problem with practical consequences. "I've been approached by people working on the neuroscience of law," Miller says. "They're trying to make inferences they can use in a court of law about people's psychiatric condition and mental states. They want to use brain scans to decide if someone should go to jail or not, so it most definitely matters if there's a systematic difference between the individual brain and the 'average' brain."[7]

Miller is not the only scientist to confront a field-shaking dilemma involving the use of averages. Every discipline that studies human beings has long relied on the same core method of research: put a group of people into some experimental condition, determine their average response to the condition, then use this average to formulate a general conclusion about *all* people. Biologists embraced a theory of the average cell, oncologists advocated treatments for the average cancer, and geneticists sought to identify the average genome. Following the theories and methods of science, our schools continue to evaluate individual students by comparing them to the average student and businesses evaluate individual job applicants and employees by comparing them to the average applicant and average employee. But if there is no such thing as an average body and there is no such thing as an average brain, this leads us to a crucial question: How did our society come to place such unquestioning faith in the idea of the average person?

The untold story of how our scientists, schools, and businesses all came to embrace the misguided notion of the "Average Man" begins in 1819, at the graduation of the most important scientist you have never heard of, a young Belgian by the name of Adolphe Quetelet.

THE MATHEMATICS OF SOCIETY

Quetelet ("Kettle-Lay") was born in 1796. At age twenty-three he received the first doctorate in mathematics ever awarded by the University of Ghent. Smart and hungry for recognition, he wanted to make a name for himself like one of his heroes, Sir Isaac Newton. Quetelet marveled at the way Newton uncovered hidden laws governing the operation of the universe, extracting orderly principles out of the chaos of matter and time. Quetelet felt that his best chance for a similar achievement was in astronomy, the leading scientific discipline of his time.[8]

In the early nineteenth century, the most prominent scientific minds turned their attention to the heavens, and the greatest symbol of a nation's scientific status was the possession of a telescopic observatory. Belgium, however, did not have one. In 1823, Quetelet somehow managed to convince the Dutch government that ruled Belgium to shell out the exorbitant sum needed to build an observatory in Brussels, and very soon Quetelet was appointed to its top position, director of the observatory.[9] As the lengthy construction proceeded, Quetelet embarked on a series of visits to observatories throughout Europe to learn the latest observational methods. It seemed he had perfectly positioned himself to make an enviable run at scientific acclaim—but then, in 1830, just as he was wrapping up his tour of Europe, Quetelet received bad news: Belgium had plunged into revolution. The Brussels observatory was occupied by rebel troops.[10]

Quetelet had no idea how long the revolution would last, or whether the new government would support the completion of the observatory—or if it would even allow him to continue as Belgium's "Astronomer Royale." It would prove to be a turning point in his life—and in the way society conceived of individuals.[11]

Previously, Quetelet never cared much about politics or the complexities of interpersonal dynamics. He was solely focused on astronomy. He believed he could keep his distance from any social commotion, which he viewed as irrelevant to his lofty scientific endeavors. But when revolution erupted in his own backyard—in his own *observatory*—human social behavior suddenly became very personal. Quetelet found himself longing for a stable government that passed sensible laws and policies that would prevent the sort of social chaos that had derailed his career plans—and which seemed to keep leading to upheaval all around Europe. There was just one glaring problem: modern society seemed utterly unpredictable. Human behavior did not appear to follow any discernible rules . . . just like the universe had seemed so indecipherable before Isaac Newton.[12]

As he contemplated the revolution that had put an end to his professional ambitions, Quetelet was struck with inspiration: Might it be possible to develop a *science* for managing society? He had spent his life learning how to identify hidden patterns in the mysterious whirl of the celestial heavens. Couldn't he use the same science to find hidden patterns in the apparent chaos of social behavior? Quetelet set himself a new goal. He would apply the methods of astronomy to the study of people. He would become the Isaac Newton of *social* physics.[13]

Fortunately for Quetelet, his decision to study social behavior came during a propitious moment in history. Europe was awash in the first tidal wave of "big data" in history, what one historian calls "an avalanche of printed numbers."[14] As nations started developing large-scale bureaucracies and militaries in the early nineteenth century, they began tabulating and publishing huge amounts of data about their citizenry, such as the number of births and deaths each month, the number of criminals incarcerated each year, and the number of incidences of disease in each city.[15] This was the very inception of modern data collection, but nobody knew how to usefully interpret this hodgepodge of data. Most scientists of the time believed that human data was far too messy to analyze—until Quetelet decided to apply the mathematics of astronomy.

Quetelet knew that one common task for any eighteenth-century astronomer was to measure the speed of celestial objects. This task was accomplished by recording the length of time it took an object such as a planet, comet, or star to pass between two parallel lines etched onto the telescope glass. For example, if an astronomer wanted to calculate the speed of Saturn and make predictions about where it would appear in the future, he would start his pocket watch when he observed Saturn touch the first line, then stop the watch when it touched the second line.[16]

Astronomers quickly discovered this technique suffered from one major problem: if ten astronomers each attempted to measure the speed of the same object, they often obtained ten different measurements. If multiple observations resulted in multiple outcomes, how could scientists decide which one to use? Eventually, astronomers adopted an ingenious solution that was originally known as the "method of averages"[17]: all the individual measurements were combined together into a single "average measurement" which, according to the advocates of the method, more accurately estimated the true value of the measurement in question than any single observation.[18]

When Quetelet ventured to establish a social science, his most pivotal decision was borrowing astronomy's method of averages and applying it to people. His decision would lead to a revolution in the way society thought of the individual.

THE AVERAGE MAN

In the early 1840s, Quetelet analyzed a data set published in an Edinburgh medical journal that listed the chest circumference, in inches, of 5,738 Scottish soldiers. This was one of the most important if uncelebrated studies of human beings in the annals of science. Quetelet added together each of the measurements, then divided the sum by the total number of soldiers. The result came out to just over thirty-nine and three-quarters inches—the average chest circumference of a Scottish soldier. This number represented one of the very first times a scientist had calculated the average of *any* human feature.[19] But it was not Quetelet's arithmetic that was history making, it was his answer to a rather simple-seeming question: What, precisely, did this average actually *mean*?

If you spend a few moments thinking about it, it's not actually clear what the significance of "average size" is. Is it a rough guide to the size of normal human beings? An estimate of the size of a randomly selected person? Or is there some kind of deeper fundamental meaning behind the number? Quetelet's own interpretation— the first scientific interpretation of a human average—was, not surprisingly, conceived out of concepts from astronomical observation.

Astronomers believed that every individual measurement of a celestial object (such as one scientist's measurement of the speed of Saturn) always contained some amount of error, yet the *total* amount of aggregate error across a group of individual measurements (such as many different scientists' measurements of the speed of Saturn, or many different measurements by a single scientist) could be minimized by using the average measurement.[20] In fact, a celebrated proof by the famous mathematician Carl Gauss appeared to demonstrate that an average measurement was as close to a measurement's true value (such as the true speed of Saturn) as you could ever hope to get.[21] Quetelet applied the same thinking to his interpretation of human averages: he declared that the individual person was synonymous with error, while the average person represented the true human being.[22]

After Quetelet calculated the average chest circumference of Scottish soldiers, he concluded that each individual soldier's chest size represented an instance of naturally occurring "error," whereas the average chest size represented the size of the "true" soldier—a perfectly formed soldier free from any physical blemishes or disruptions, as Nature intended a soldier to be.[23] To justify this peculiar interpretation, Quetelet offered an explanatory metaphor known as the "Statue of the Gladiator."

Quetelet invites us to imagine a statue of a gladiator. Suppose that sculptors make 1,000 copies of the statue. Quetelet claims that every one of these hand-carved copies will always feature some mistakes and flaws that will render it different from the original. Yet, according to Quetelet, if you took the average of all 1,000 copies, this "average statue" would be nearly identical to the original statue. In the same manner, contended Quetelet in a striking leap of logic, if you averaged together 1,000 different soldiers, you would end up with a very close approximation of the One True Soldier, existing in some Platonic realm, of which each living, breathing soldier was an imperfect representation.[24]

Quetelet followed the same line of reasoning with regard to humanity as a whole, claiming that every one of us is a flawed copy of some kind of cosmic template for human beings. Quetelet dubbed this template the "Average Man."[25] Today, of course, we often consider someone described as "average" to be inferior or lacking—as mediocre. But for Quetelet, the Average Man was perfection itself, an ideal that Nature aspired to, free from Error with a capital "E." He declared that the greatest men in history were closest to the Average Man of their place and time.[26]

Eager to unmask the secret face of the Average Man, Quetelet began to compute the average of every human attribute he could get data on. He calculated average stature, average weight, and average complexion. He calculated the average age couples got married and the average age people died. He calculated average annual births, average number of people in poverty, average annual incidents of crime, average types of crimes, the average amount of education, and even average annual suicide rates. He invented the Quetelet

Index—today known as the body mass index (BMI)—and calcu-lated men's and women's average BMIs to identify average health. Each of these average values, claimed Quetelet, represented the hid-den qualities of the One True Human, the Average Man.

As much as Quetelet admired the Average Man, he held an equal amount of antipathy toward those unfortunate individuals who deviated from the average. "Everything differing from the Aver-age Man's proportions and condition, would constitute deformity and disease," Quetelet asserted. "Everything found dissimilar, not only as regarded proportion or form, but as exceeding the observed limits, would constitute a Monstrosity."[27] There's little question that Quetelet would have lauded the statue of Norma. "If an individual at any given epoch of society possessed all the qualities of the Aver-age Man," pronounced Quetelet, "he would represent all that is great, good, or beautiful."[28]

Though today we don't think an average person is perfection, we do presume that an average person is a prototypical representa-tive of a group—a *type*. There is a powerful tendency in the human mind to simplify the way we think about people by imagining that all members of a group—such as "lawyers," "the homeless," or "Mexicans"—act according to a set of shared characteristics, and Quetelet's research endowed this impulse with a scientific justifica-tion that quickly became a cornerstone of the social sciences. Ever since Quetelet introduced the idea of the Average Man, scientists have delineated the characteristics of a seemingly endless number of types, such as "Type-A personalities," "neurotic types," "micro-managers," and "leader types," arguing that you could make useful predictions about any given individual member of a group simply by knowing the traits of the average member—the group's type.

Since Quetelet's new science of the Average Man seemed to impose welcome order on the accelerating jumble of human statistics while simultaneously validating people's natural urge to stereotype others, it's little wonder his ideas spread like wildfire. Governments adopted Quetelet's social physics as a basis for understanding their citizens and crafting social policy. His ideas helped focus political attention on the middle class, since they were perceived to be closest to a nation's average citizen and, according to Queteletian reasoning, the truest type of Belgian, Frenchman, Englishman, Dutchman, or Prussian. In 1846, Quetelet organized the first census for the Belgian government, which became the gold standard for all modern censuses; Quetelet even consulted with James A. Garfield, then a member of the U.S. Congress, about ways to improve the American census.[29]

Quetelet also influenced the American military. During the American Civil War, President Abraham Lincoln decided that the Union Army needed more information about its soldiers in order to provide a more efficient distribution of resources, so he authorized the largest anthropometric study in the history of the world up to that time. Every Union soldier was measured physically, medically, and morally, and then—in explicit obedience to Quetelet's new science—averages were calculated and reported. This mammoth study formed the basis for the American military's long-standing philosophy of standardized design.[30]

You and I take averages for granted. They form part of the everyday babble and hum of our daily media. As I write this, today's *New York Times* reports the average amount of student debt, the average number of viewers of prime-time television, and the average salary of physicians. But each time Quetelet unveiled a new average, the public boggled. For example, Quetelet showed that the average rate of suicide was relatively stable from year to year.[31] While this would hardly be startling news for us, in the 1830s suicide seemed to be a highly irrational private decision that could not possibly conform to any deeper pattern. Instead, Quetelet showed that suicides occurred with reliable and consistent regularity—and not only that, he claimed that the stability of the occurrences indicated that everyone possesses an average propensity toward suicide. The Average Man, attested Quetelet, was suicidal to an average extent.[32]

Scholars and thinkers in every field hailed Quetelet as a genius for uncovering the hidden laws governing society. Florence Nightingale adopted his ideas in nursing, declaring that the Average Man was "God's Will." Karl Marx adopted Quetelet's ideas to develop his economic theory of Communism, announcing that the Average Man proved the existence of historical determinism. The physicist James Maxwell was inspired by Quetelet's mathematics to formulate the classical theory of gas mechanics. The physician John Snow used Quetelet's ideas to fight cholera in London, marking the start of the field of public health. Wilhelm Wundt, the father of experimental psychology, read Quetelet and proclaimed, "It can be stated without exaggeration that more psychology can be learned from statistical averages than from all philosophers, except Aristotle."[33]

Quetelet's invention of the Average Man marked the beginning of the Age of Average. It represented the moment when the average became normal, the individual became error, and stereotypes were validated with the imprint of science. These assumptions would eventually lead the Air Force to design cockpits to fit the average pilot and my instructors at Mass General Hospital to teach me how to interpret maps of the Average Brain. It would prompt generations of parents to worry if their child did not develop according to the average milestones, and cause almost every one of us to feel anxiety when our health, social life, or career deviated too far from the average.

But Quetelet is only half the story of how the Age of Average came about. The other half centers on Sir Francis Galton, a giant of a figure who started out as one of Quetelet's most devout disciples but eventually became his most distinguished detractor.[34]

THE EMINENT AND THE IMBECILE

In 1851, the Great Exhibition—sometimes called the first World's Fair—was held in London. Exhibitors from every nation showcased their most interesting products, technologies, and inventions. The British people fully expected the event would demonstrate to the world their country's superiority. But as they strolled through the exhibits, it quickly became apparent that their hopes were not being fulfilled. The most impressive exhibits were not British, but American. Entrepreneurs from across the Atlantic touted industrial marvels that surpassed anything the British had on offer, including Samuel Colt's revolver, Isaac Singer's sewing machine, and Robert McCormick's mechanical reaper.[35] Many Englishmen began to worry their country was falling behind the rest of the world—and one man who was especially concerned was Francis Galton. He was sure he knew the precise cause of the United Kingdom's abrupt downturn: the growing status of the lower classes.[36]

Galton, whose family had made its fortune in banking and gun manufacturing, was a member of the wealthy merchant class. He believed in the innate superiority of his family and other members of the upper class, and to his mind, the growing democratization of society was polluting the greatness of the British Empire.[37] He was confident that the way to restore Britain's tarnished glory was to reestablish the lapsing authority of the superior social strata—and he believed Quetelet's math explained why.

A mathematician by training, Galton viewed the elder Belgian as brilliant, calling him "the greatest authority on vital and social statistics."[38] Galton concurred with Quetelet that the average represented the scientific foundation for understanding people. In fact, Galton agreed with almost all of Quetelet's ideas, save one: the idea that the Average Man represented Nature's ideal. Nothing could be further from the truth, claimed Galton. For him, to be average was to be mediocre, crude, and undistinguished—like the lower classes who were now voting for representatives in the House of Commons.[39] Galton would have scoffed at the idea that women should try to fashion themselves after Norma. No, if women wanted a model to emulate, Galton believed there was none better than Her Majesty the Queen.

Galton believed that it was the imperative of humankind to attempt to *improve* on the average as much as possible, and he cited his cousin Charles Darwin's research to support this claim, writing, "What nature does blindly, slowly, and ruthlessly, man may do providently, quickly, and kindly."[40] Though Quetelet considered that excessive deviation from the average constituted "monstrosity," Galton believed the Belgian's view was only half right. Luminaries who were far above average—like Galton and Queen Victoria and Isaac Newton—were assuredly not monstrosities, but instead formed a distinct class that Galton dubbed "the Eminent." Those who were far below average Galton termed "the Imbecile."[41]

Thus, Galton rejected Quetelet's conviction that individuals who deviated from the average represented "error." At the same time, he agreed with Quetelet's concept of types, since he believed that the Eminent, the Imbecile, and the Mediocre each comprised a separate type of human being. Put simply, Galton wanted to preserve Quetelet's idea that the average member of a group represented that group's type, but reject Quetelet's idea that an individual's deviation from average represented error. How did he resolve this apparent paradox? Through an act of moral and mathematical jujitsu: he redefined "error" as "rank."[42]

Quetelet might say that it did not really matter whether you were 50 percent faster than the average person or 50 percent slower—in either case, you were an equal deviation from the average, embodying equal error and equal distance from perfection. Galton would have disagreed. He said that a person who was 50 percent faster than average was clearly *superior* to someone 50 percent slower. They were not equal: the faster person represented an individual of higher *rank*.

Galton carved up humankind into fourteen distinct classes, ranging from the "Imbeciles" in the lowest rank through the "Mediocre" in the middle ranks all the way up to the most "Eminent" members of the highest rank. This was a monumental shift in the meaning of average, transforming the notion of normality into mediocrity. But Galton didn't stop there. Galton was so confident that the Eminent represented a separate category of human being that he claimed a person's rank was consistent across all qualities and dimensions—mental, physical, and moral.[43] According to Galton, if your intelligence was Eminent, your physical health would most likely be Eminent, too, as well as your courage and honesty. Similarly, if your math skills lingered in the lowest ranks, your verbal skills would probably also slouch far below average, not to mention your beauty and self-discipline. "As statistics have shown, the best qualities are largely correlated," wrote Galton in 1909.[44] "The youths who became judges, bishops, statesman, and leaders of progress in England could have furnished formidable athletic teams in their times."

If Galton's conception of rank were true it would support his contention that the Eminent offered the best hope for restoring Britain's lapsed glory, since it would mean the Eminent as a class were eminent in all things. To help him prove the existence of rank, Galton developed new statistical methods, including *correlation,* a technique that allowed him to assess the relationship of rank across different qualities.

All of his statistical inventions were predicated on what Galton called the "law of deviation from the average": the idea that what mattered most about an individual was how much better or worse they were than the average. To our twenty-first-century minds, it has come to seem so natural and obvious that talented people are "above average" while incompetent folks are "below average" that it seems simplistic to attribute the origins of this idea to one person. And yet, it was Galton who almost single-handedly supplanted Quetelet's conviction that human worth could be measured by how close a person was to the average with the notion that worth was better measured by how *far* a person was from the average. Just as Quetelet's ideas about types took the intellectual world by storm in the 1840s, so did Galton's idea about rank in the 1890s, and by the early 1900s, the notion that people could be sorted into distinct bins of ability from low to high had infiltrated virtually all the social and behavioral sciences.

The Age of Average—a cultural era stretching from Quetelet's invention of social physics in the 1840s until today—can be characterized by two assumptions unconsciously shared by almost every member of society: Quetelet's idea of the average man and Galton's idea of rank. We have all come to believe, like Quetelet, that the average is a reliable index of normality, particularly when it comes to physical health, mental health, personality, and economic status. We have also come to believe that an individual's rank on narrow metrics of achievement can be used to judge their talent. These two ideas serve as the organizing principles behind our current system of education, the vast majority of hiring practices, and most employee performance evaluation systems worldwide.

Though Quetelet's influence on the way we think about individuals remains deeply embedded in our institutions, for most of us, it is Galton's legacy that clenches itself around our personal life in a more vivid and intimate manner. We all feel the pressure to strive to rise as far above average as possible. Much of the time, we don't even think about *what,* exactly, we're trying so hard to be above-average at, because the *why* is so clear: we can only achieve success in the Age of Average if others do not view us as mediocre or— disaster!—as below-average.

THE RISE OF THE AVERAGARIANS

By the dawn of the twentieth century, a majority of social scientists and policymakers were making decisions about people based on the average.[45] This development did not merely consist of the adoption of new statistical techniques. It marked a seismic change in how we conceived of the relationship between the individual and society. Typing and ranking both rely on a comparison of the individual to a group average. Thus, both Quetelet and Galton claimed, explicitly and ardently, that any particular person could only be understood by comparison to the group, and therefore, from the perspective of the new social sciences, the individual was almost entirely irrelevant.

"In speaking of the individual it must be understood that we are not attempting to speak of this or that man in particular; we must turn to the general impression that remains after having considered a great number of people," Quetelet wrote in 1835. "Removing his individuality we will eliminate all that is accidental."[46] Similarly, the first issue of *Biometrika,* an academic journal that Galton founded in 1901, proclaimed, "It is almost impossible to study any type of life without being impressed by the small importance of the individual."[47] It might seem that there is some fundamental difference between saying a person scored in the 90th percentile and saying that a person is an introverted type, but both ultimately require a comparison to an average score. These two approaches merely reflect an alternate interpretation of the same underlying mathematics— but share the same core conviction: individuality doesn't matter.

When the average was first introduced into society, many educated Victorians recognized right away that something vital was under threat by their strange new approach to understanding people, driving many to warn, rather prophetically, of the perils of ignoring individuality. In an 1864 essay, a well-known British poet named William Cyples acknowledged the ostensible triumphs of a new generation of average-wielding scientists and bureaucrats, before endowing them with a moniker as distinctive as it was disparaging: *averagarians*. This term is so useful and apt that I employ it to describe anyone—scientists, educators, managers—who use averages to understand individuals.

In his essay, Cyples worried about what the future would look like if the averagarians took over: "These averagarians usually give the statistics of murders, suicides, and (unhappy connection!) marriages, as proof of the periodic uniformity of events. . . . We should seem rather to be human units than men. . . . We endure or achieve in the degree of a percentage; fate is not so much a personal ordainment as an allotment made on us in statistical groups. . . . A protest may be safely entered against this modern superstition of arithmetic which, if acquiesced in, would seem to threaten mankind with a later and worse blight than any it has yet suffered—that not so much of a fixed destiny, as of a fate expressive in decimal fractions, not falling upon us personally, but in averages."[48]

It wasn't just poets who were concerned about the growing influence of the averagarians. Physicians, too, were staunchly opposed to the use of the average to evaluate individuals under their care. "You can tell your patient that, of every hundred such cases, eighty are cured . . . but that will scarcely move him. What he wants to know is whether he is numbered among those who are cured," wrote Claude Bernard, the French doctor regarded as the father of experimental medicine, in 1865.[49] "Physicians have nothing to do with what is called the law of large numbers, a law which, according to a great mathematician's expression, is always true in general and false in particular."[50]

Yet society failed to listen to these early protests, and today we reflexively judge every individual we meet in comparison to the average—including ourselves. When the media reports the number of close friends the average citizen possesses (8.6 in the United

States), or the number of romantic partners the average person kisses in a lifetime (15 for women, 16 for men), or the number of fights over money the average couple instigates each month (3 in the United States)—it is the rare person who doesn't automatically weigh her own life against these figures. If we have claimed more than our fair share of kisses, we may even feel a surge of pride; if we have fallen short, we may feel self-pity or shame.[51]

Typing and ranking have come to seem so elementary, natural, and right that we are no longer conscious of the fact that every such judgment always erases the individuality of the person being judged. A century and a half after Quetelet—exactly as the poets and physicians of the nineteenth century feared—we have all become averagarians.

CHAPTER 2

HOW OUR WORLD BECAME STANDARDIZED

After I dropped out of high school, I worked briefly for a large aluminum stamping plant in Clearfield, Utah. It served as my introduction to the world of work. On my first day, I was handed a little card that explained in precise detail exactly how I was supposed to do my job, even dictating the preferred motions of my arms and feet. I picked up a raw block of aluminum from a pile and carried it over to the blistering hot stamping machine. I ran the block through the machine, which squeezed out an L- or S-shaped beam like a Play-Doh Fun Factory. I stacked the beam on a pallet and punched a button recording the fact that I had completed one unit (a portion of my pay was based on the number of units I stamped), then I raced back to the pile and started all over again.

My two most enduring memories of the job are the endless repetition of punching and running, punching and running, punching and running . . . and the piercing metallic clanging of the factory bell announcing the start and end of my shift. The experience was dehumanizing. As an aluminum plant employee, my individuality did not matter at all. Instead, just as the British poet Cyples had warned, I was a "human unit"—a mere statistic, an Average Worker. This was no coincidence: the entire workplace was designed according to the tenets of averagarianism, the supposition that individuals can be evaluated, sorted, and managed by comparing them to the average.

Since averagarianism originally grew out of the work of two European scientists who were trying to use mathematics to solve complex social problems, it could have remained an esoteric philosophy of interest only to scholars or intellectuals. But you and I were born into a world where the notion of the average colors every aspect of our lives from birth to death, infiltrating our most confidential judgments of self-worth. How, precisely, did averagarianism go from an abstract ivory tower conjecture to the pre-eminent organizational doctrine of businesses and schools across the world? The answer to this question largely centers on a single man named Frederick Winslow Taylor.

One economist has written that Taylor "probably had a greater effect on the private and public lives of the men and women of the twentieth century than any other single individual."[1] Born to a wealthy Pennsylvania family in 1856, the teenage Taylor spent two years studying in Prussia, one of the first countries to re-organize its schools and military around Quetelet's ideas, and the place where Taylor was most likely first exposed to the ideas of averagarianism that would eventually form the philosophical foundation of his work.[2]

After Taylor returned home to attend Phillips Exeter Academy, a college prep school, his family expected him to follow in his father's footsteps and study law at Harvard. Instead, he became an apprentice machinist at a Philadelphia pump-manufacturing company.

When I first read about Taylor's youthful career decision, I thought I recognized a kindred spirit, imagining Taylor as a troubled adolescent who couldn't find his way through school or life. I was mistaken. Taylor's decision to work for a pump manufacturer is better compared to an ambitious Mark Zuckerberg dropping out of Harvard to found Facebook.

In the 1880s, America was transitioning from an agrarian economy to an industrial one. Newly laid railways linked cities in webs of iron, immigrants flooded into the country so fast that you could walk through entire neighborhoods without hearing a word of English, and cities expanded so rapidly that between 1870 and 1900 the population of Chicago increased sixfold. These social disruptions were accompanied by significant economic changes, and the biggest ones were taking place inside the giant new edifices of manufacturing: factories. Taylor skipped Harvard to work at Enterprise Hydraulic Works at the dawn of electrified factories, a time when making things, assembling things, and building things roused the same sense of world-conquering opportunity found today in Silicon Valley.[3]

Taylor hoped to make a name for himself in this thrilling new world of industry, an ambition much aided by the fact that the pump company was owned by family friends. Taylor had to do a bare minimum of heavy labor, so he was largely free to observe and contemplate the details of the factory's operations. When he finished his apprenticeship he became a machine-shop laborer at Midvale Steelwork, another friend-owned business, where he was put on the fast track to promotions. After six promotions in six years, he was appointed chief engineer for the entire company.[4]

During these six years, Taylor contemplated the problems of the new era of factory production. There were plenty. The early decades of the Second Industrial Revolution were characterized by runaway inflation, plunging wages, and frequent financial panics. When Taylor first started at Midvale, the country was in the midst of the worst depression of his generation. Workers rarely stayed in one place, and factories had turnover rates ranging from 100 percent to 1,500 percent in a single year.[5] Nobody really understood what was causing all the new economic problems of the factory age, but by the time he became chief engineer, Taylor was convinced that he knew their true source: inefficiency.[6]

The new electrified factories, Taylor asserted, wasted outrageous amounts of labor. All of this waste was the result of the way factories organized its workers, which, to Taylor's mind, was clumsy, inept, and—most importantly—unscientific. Seven decades earlier, at the dawn of the Industrial Revolution, the first large-scale industries of textile manufacturing, iron making, and steam power created massive social upheaval, prompting Adolphe Quetelet to try to solve these new social problems through a science of society. Quetelet became the Isaac Newton of social physics. Now, in the 1890s, Taylor looked at a new era of economic upheaval and declared that the problems of the factory age could only be solved through a science of work. In other words, Taylor set out to become the Adolphe Quetelet of industrial organization.

He believed that he could systematically eliminate inefficiency from business by adopting the core precept of averagarianism, the idea that individuality did not matter. "In the past the man was first," announced Taylor, "in the future the system must be first."[7]

THE SYSTEM MUST BE FIRST

Before Taylor set out to develop a new science of work, companies usually hired the most talented workers available, regardless of their particular skill set, and then let these star employees reorganize a company's processes according to what they believed would help them be most productive. Taylor insisted this was completely back-ward. A business should not conform its system to fit individual employees, no matter how special they were perceived to be. Instead, businesses should hire Average Men who fit the system. "An orga-nization composed of individuals of mediocre ability, working in accordance with policies, plans, and procedures discovered by analy-sis of the fundamental facts of their situation, will in the long run prove more successful and stable than an organization of geniuses each led by inspiration," affirmed Taylor.[8]

Starting in the 1890s, Taylor began sharing a new vision for industrial organization that he suggested would minimize ineffi-ciency in the same way that the method of averages was presumed to minimize error. His vision was grounded on one key averagar-ian concept: standardization.[9] Though Quetelet was the first scien-tist to champion standardization in government bureaucracies and scientific data collection, Taylor said that his own inspiration for standardizing human labor came from one of his math teachers at Phillips Exeter Academy.[10]

The teacher often assigned Taylor and his classmates a series of math problems, instructing each boy to snap his fingers and raise his hand when he completed the problems. The teacher used a stop-watch to time his students, then calculated how long it took the average boy to finish. Then, when the teacher created homework assignments, he used this average time to calculate exactly how many problems he needed to include in the assignment so that it would take the average boy exactly two hours to complete.

Taylor realized his teacher's method of standardizing homework could also be used to standardize any industrial process.[11] His earli-est attempts at standardization took place at Midvale Steelworks. First, Taylor looked for ways to improve the speed of any given task in the factory, such as shoveling coal into the furnace. Once a task was optimized according to Taylor's satisfaction, he measured the average time it took workers to complete the task. He also determined the average physical motions that workers used to perform that task. For example, he determined that the optimal amount of coal to shovel in a single swing was 21 pounds. Taylor then standardized the entire industrial process around these averages so that the way to perform each task became fixed and inviolable (in the case of coal shoveling, he insisted that special shovels optimized to carry 21 pounds were always used), and workers were not permitted to deviate from these standards—just as I was required to stamp aluminum in a precisely prescribed way.

According to Taylor, there was always "one best way" to accomplish any given process—and only one way, the standardized way.[12] For Taylor, there was nothing worse than a worker trying to do things *his* own way. "There is a rock upon which many an ingenious man has stranded, that of indulging his inventive faculty," warned Taylor in a 1918 magazine article. "It is thoroughly illegitimate for the average man to start out to make a radically new machine, or method, or process to replace one which is already successful."[13] American factories embraced Taylor's principles of standardization and were soon posting work rules, printing books of standard operating procedures, and issuing job instruction cards, all laying out the requisite way to get things done. The worker, once celebrated as a creative craftsman, was relegated to the role of automaton.[14]

Today, standardization is implemented in modern enterprises in a form virtually unchanged from Taylor's earliest proposals, a form I experienced firsthand at the aluminum stamping plant. Since it was my first real full-time job, I thought its dehumanizing grind was unique to one particular company in Utah. I was quickly disabused of that notion. Two years later, I was hired as a customer service rep for a major credit card company, sitting in a comfy swivel chair in an air-conditioned office. It seemed like it would be much different from my factory job. It wasn't. My role, once again, was completely shaped by Taylor's principles of standardization.

I was given a detailed script to use on calls and instructed not to veer from this script in any way. Since following the script correctly meant that a customer service call would last an average length of time, I was evaluated on the duration of each and every call. If a call exceeded the average time, my screen began flashing red. Instead of focusing on the quality of the call, I focused on making sure I hit the disconnect button as fast as possible. The computer updated my average time after each call and showed how I compared to the group average—and shared my average with my supervisor, too. If my average exceeded the group average by too much, my supervisor paid me a visit, which he did, several times. If my average had remained high, he could have fired me—though I quit before that could actually happen.

Over the next few years I worked in retail, restaurants, sales, and factories, and in each and every organization my job was standardized according to Taylor's belief that "the system must be first." Each time, I was a cog in a machine, with no opportunity to express individual initiative or take individual accountability. Each time, I was expected to conform to the average as closely as possible—or to be like everyone else, only better. What was worse, when I complained about how these jobs failed to take my own personality into account, leaving me feeling helpless and bored, I was often accused of being lazy or irresponsible. In a standardized system, individuality does not matter, and that was exactly what Taylor intended.

THE BIRTH OF THE MANAGER

Standardization left one crucial question unanswered: Who should create the standards that governed a business? Certainly not the worker, insisted Taylor. He argued that businesses should take away all planning, control, and decision making from the workers and hand it over to a new class of "planners" who would be responsible for overseeing the workers and determining the one best way to standardize an organization's processes. Taylor adopted a recently invented term to describe this new role: "the manager."[15]

While the notion of managers may seem like a fairly obvious idea to our modern minds, it ran counter to the conventional wisdom of nineteenth-century business. Before Taylor, companies viewed "nonproductive" employees who sat at a desk without doing physical labor as an unnecessary expense. It didn't seem to make any sense to hire someone to plan a job who couldn't actually *do* the job. But Taylor insisted this view was all wrong. Factories needed brains to direct the hands.[16] It needed planners to figure out the one best way to set up the stamping machines, the one best way to stamp the aluminum, and the one best way to hire, schedule, pay, and fire workers. It was Taylor's singular vision that shaped our modern sense of the manager as an executive decision-maker.

Taylor also established the fundamental division of business roles that quickly came to define our modern workplace: the managers in charge of running the show, and the employees who actually did the work. In Taylor's time, these employees were primarily factory workers, but today they include roles as varied as administrative assistants, phlebotomists, air traffic controllers, electrical engineers, and pharmaceutical researchers. At a 1906 lecture, Taylor explained how he saw the relationship between workers and managers: "In our scheme, we do not ask for the initiative of our men. We do not want any initiative. All we want of them is to obey the orders we give them, do what we say, and do it quick."[17] In 1918, Taylor doubled down on these ideas, dishing out similar advice to aspiring mechanical engineers: "Every day, year in and year out, each man should ask himself, over and over again, two questions: First, 'What is the name of the man I am now working for?' and . . . 'What does this man want me to do?' The most important idea should be that of serving the man who is over you his way, not yours."[18]

Taylor laid out his ideas of standardization and management in his 1911 book *The Principles of Scientific Management*.[19] The book became a national and international business bestseller and was translated into a dozen languages.[20] Almost immediately after the book's publication, scientific management—often simply called "Taylorism"—swept across the world's industries.

Business owners restructured their enterprises by creating departments and subdepartments, each headed by a Taylorist manager, making the organizational chart ("org chart") a new focal point. Personnel and human resources departments were established and tasked with finding and hiring employees and assigning them to jobs. Taylorism inaugurated planning departments, efficiency experts, industrial-organizational psychology, and time-study engineering. (A single Westinghouse plant in 1929 had a time-study staff of a hundred and twenty who set the standards for more than a hundred thousand industrial processes each month.)[21]

Since thinking and planning were now cleanly separated from making and doing, businesses developed an insatiable appetite for experts to tell them the best way to do all that thinking and planning. The management consulting industry was born to satisfy this appetite, and Frederick Taylor became the world's first management consultant. His opinion was so highly sought after that he sometimes charged the modern equivalent of $2.5 million for his advice.

All these management consultants, planning departments, and efficiency experts relied on the mathematics of the average to conduct their analyses. Managers believed that the science of Quetelet and Galton justified treating each worker like a cell on a spreadsheet, as a number in a column, an interchangeable Average Man.

It was not very difficult to convince managers that individuality did not matter, since it made their job easier and more secure. After all, if you make decisions about people using types and ranks, you will not be right every time—but you will tend to be right on average, and that was good enough for large organizations with many standardized processes and roles. On those occasions where managers did make a wrong decision about an employee, they could simply blame the employee for not fitting into the system.

United States Rubber Company, International Harvester Company, and General Motors were all early adopters of the principles of scientific management. Taylorism was also applied to bricklaying, canning, food processing, dyeing, bookbinding, publishing, lithography, and wire weaving, and then to dentistry, banking, and the manufacturing of hotel furniture. In France, Renault applied Taylorism to automaking and Michelin applied it to the manufacture of tires. President Franklin Roosevelt's system of national planning was explicitly modeled on Taylorism. By 1927, scientific management had already become so widely adopted that a League of Nations report called it "a characteristic feature of American civilization."[22]

Even though Taylorism was often equated with American capitalism, its appeal crossed borders and ideologies. In Soviet Russia, Lenin heralded scientific management as the key to jump-starting Russian factories and organizing five-year industrial plans, and by the start of World War II, Frederick Taylor was as famous in the Soviet Union as Franklin Roosevelt. Mussolini and Hitler added their names to Lenin and Stalin as ardent supporters of Taylorism, adopting it for their war industries.[23]

Meanwhile, collectivist cultures in Asia applied scientific management even more ruthlessly than their Western counterparts, with companies like Mitsubishi and Toshiba completely remaking themselves according to the principles of standardization and worker-manager separation. When Taylor's son visited Japan in 1961,

Toshiba executives begged him for a pencil, a picture, anything that had been touched by his father.[24]

Today, scientific management remains the most dominant philosophy of business organization in every industrialized country.[25] No company likes to admit it, of course, since in many circles Taylorism has acquired the same disreputable connotation as racism or sexism. But many of the largest and most successful corporations on Earth are still organized around the idea that the individuality of the employee does not matter.

All of this leads to a profound question that transcends Taylorism: If you have a society predicated upon the separation of system-conforming workers from system-defining managers, how does society decide who gets to be a worker and who gets to be a manager?

FACTORIES OF EDUCATION

As Taylorism began to transform American industry at the dawn of the twentieth century, factories began to develop an insatiable need for semi-skilled workers who possessed a high school education. But there was a problem. Not only did the country lack universal high school education, there were hardly any high schools at all. In the year 1900, roughly 6 percent of the American population graduated from high school. Just 2 percent graduated from college.[26] At the same time, there was a massive influx of children of immigrants and factory workers, particularly in the cities, threatening to increase the number of uneducated youth still further. It soon became apparent to everyone that the American educational system needed a major overhaul.

The question that occupied the earliest education reformers was what the mission of the new school system should be. A group of educators with a humanist perspective argued that the proper goal of education was to provide students with the freedom to discover their own talents and interests by offering an environment that would allow them to learn and develop at their own pace. Some humanists even suggested that there should be no required courses, and that schools should offer more courses than any student could possibly take.[27] But when it came time to establish a nationwide, compulsory high school system, the humanist model was passed over in favor of a very different vision of education—a Taylorist vision.

It was never a fair fight. On one side stood the humanists, a coterie of tweed-coated academics at cushy, exclusive northeastern colleges. They were opposed by a broad coalition of pragmatic industrialists and ambitious psychologists steeped in the values of standardization and hierarchical management. These educational Taylorists pointed out that while it was nice to think about humanistic ideals like educational self-determination, at a time when many public schools had a hundred kids in a single classroom, half unable to speak English, many living in poverty, educators did not have the luxury of giving young people the freedom to be whatever they wanted to be.[28]

The educational Taylorists declared that the new mission of education should be to prepare mass numbers of students to work in the newly Taylorized economy. Following Taylor's maxim that a system of average workers was more efficient than a system of geniuses, educational Taylorists argued that schools should provide a standard education for an average student instead of trying to foster greatness. By way of example, John D. Rockefeller funded an organization known as the General Education Board, which published a 1912 essay describing its Taylorist vision of schools: "We shall not try to make these people or any of their children into philosophers or men of learning or of science. We are not to raise up from among them authors, orators, poets, or men of letters. We shall not search for embryo great artists, painters, musicians . . . nor lawyers, doc-

tors, preachers, politicians, statesmen, of whom we have ample supply. . . . The task that we set before ourselves is very simple as well as
very beautiful . . . we will organize our children into a little community and teach them to do in a perfect way the things their fathers
and mothers are doing in an imperfect way." [29]

To organize and teach children to become workers who could
perform industrial tasks in "a perfect way," the Taylorists set out to
remake the architecture of the entire educational system to conform
to the central tenet of scientific management: standardize everything around the average. Schools around the country adopted the
"Gary Plan," named after the industrialized Indiana city where it
originated: students were divided into groups by age (not by performance, interest, or aptitude) and these groups of students rotated
through different classes, each lasting a standardized period of time.
School bells were introduced to emulate factory bells, in order to
mentally prepare children for their future careers. [30]

The Taylorist educational reformers also introduced a new professional role into education: the curriculum planner. Modeled after
scientific management, these planners created a fixed, inviolable curriculum that dictated everything that happened in school, including
what and how students were taught, what textbooks should contain,
and how students were graded. As standardization spread through
schools nationwide, school boards rapidly adopted a top-down hierarchical management that replicated the management structure of
Taylorism, assigning executive planning roles to principals, superintendents, and district superintendents.

By 1920, most American schools were organized according to the
Taylorist vision of education, treating each student as an average
student and aiming to provide each one with the same standardized
education, regardless of their background, abilities, or interests. In
1924, the American journalist H. L. Mencken summarized the state
of the educational system: "The aim of public education is not to

spread enlightenment at all; it is simply to reduce as many individuals as possible to the same safe level, to breed and train a standardized citizenry, to put down dissent and originality. That is its aim in
the United States . . . and that is its aim everywhere else."[31]

American schools, in other words, were staunchly Queteletian,
their curriculum and classrooms designed to serve the Average Student and create Average Workers. Even so, one man felt that the
educational Taylorists had not taken averagarianism far enough. In
an eerie parallel, just as Galton had once embraced Quetelet's ideas
about the Average Man before retrofitting the elder Belgian's ideas
so that Galton could use them to separate society's superior citizens
from its inferior ones, Edward Thorndike embraced Taylor's ideas
about standardization before refitting the elder American's ideas so
that Thorndike could use them to separate school's superior students
from its inferior ones.

THE GIFTED AND THE USELESS

Thorndike was one of the most prolific and influential psychologists of all time.[32] He published more than four hundred articles,
and sold millions of textbooks.[33] His mentor at Harvard, William
James, described Thorndike as a "freak of nature" for his workaholic
productivity. He helped invent the fields of educational psychology
and educational psychometrics as he pursued his most influential
achievement of all: establishing the mission of schools, colleges, and
universities in the Age of Average.

Thorndike fully supported the Taylorization of schools. In fact,
Thorndike played a leading role in the country's largest training program for school superintendents, preparing them for their roles as
scientific managers in the standardized educational system.[34] But
Thorndike believed that Taylorists were making a mistake when

they argued that the goal of education was to provide every student with the same average education to prepare them for the same average jobs. Thorndike believed that schools should instead sort young people according to their ability so they could efficiently be appointed to their proper station in life, whether manager or worker, eminent leader or disposable outcast—and so that educational resources could be allocated accordingly. Thorndike's guiding axiom was "Quality is more important than equality," by which he meant that it was more important to identify superior students and shower them with support than it was to provide every student with the same educational opportunities.

Thorndike was an enthusiastic advocate of the ideas of Francis Galton, whom he revered as "an eminently fair scientific man."[35] He agreed with Galton's notion of rank, the theory that if a person was talented at one thing, he was likely to be talented at most other things, too. He justified this conviction using his own biological theory of learning: Thorndike believed that some people were simply born with brains that learned quickly, and these fast-learning individuals would not only be successful at school, they would be successful in life. On the other hand, some people were born with slow brains; these poor souls were destined to fare poorly at school and would struggle all life long.

Thorndike believed that schools should clear a path for talented students to proceed to college, and then onward into jobs where their superior abilities could be put to use leading the country. The bulk of students, whose talents Thorndike assumed would hover around the average, could go straight from high school graduation—or even earlier—into their jobs as Taylorist workers in the industrial economy. As to the slow-learning students, well . . . Thorndike thought we should probably stop spending resources on them as soon as possible.[36]

So how, precisely, should schools go about ranking students?

Thorndike answered this question in his book ironically titled *Individuality,* where he redefined individuality according to the Galtonian definition: that a person's uniqueness and value stemmed from his deviation from the average.[37] Thorndike agreed that every aspect of the educational system should be standardized around the average, not only because this would ensure standardized outcomes, as the Taylorists believed, but because it made it easier to measure each student's deviation from the average—and thus made it easier to determine who was superior and who was inferior.

To help establish his desired system of student ranking, Thorndike created standardized tests for handwriting, spelling, arithmetic, English comprehension, drawing, and reading, all of which were quickly adopted by schools across the country.[38] He wrote textbooks for arithmetic, vocabulary, and spelling that were all standardized around the average student of a particular age, a practice still used in our school systems today. He designed entrance exams for private schools and elite colleges; he even fashioned an entrance exam for law school.[39] Thorndike's ideas gave birth to the notion of gifted students, honors students, special needs students, and educational tracks. He supported the use of grades as a convenient metric for ranking students' overall talent and believed that colleges should admit those students with the best GPAs and highest standardized test scores since (according to Galton's idea of rank) he believed they were not only the most likely to succeed in college, but most likely to succeed in whatever profession they chose.

For Thorndike, the purpose of schools was not to educate *all* students to the same level, but to *sort* them, according to their innate level of talent. It is deeply ironic that one of the most influential people in the history of education believed that education could do little to change a student's abilities and was therefore limited to identifying those students born with a superior brain—and those born with an inferior one.

Like so many other students, I felt the full weight of Thorndikian rankings on my aspirations for the future. In high school I took a standardized college aptitude test that is widely used as an admissions criterion by most American universities. Thorndike would have loved the test because not only does it report your ranking, the test uses this ranking to predict how you will perform at different colleges, should you choose to attend. I've tried to forget everything about my test results, but memory traces still endure like the painful residue of a traumatic experience. My score placed me in the area that Galton would have termed "Mediocrity," and the test informed me that, based on this score, the probability of me getting a B or higher at Weber State University, an open-enrollment school in Ogden, Utah, was a disheartening 40 percent. But that was still better than the odds of me getting a B or higher at my top choice, Brigham Young University: a mere 20 percent.

I remember reading these predictions and feeling pretty hopeless about my life. After all, these percentages, arranged in tidy columns, were endowed with the sober authority of mathematics: I felt like this single test had weighed my entire worth as a person and found me wanting. I initially thought I might one day be an engineer or a neurologist, but no—what a silly fantasy *that* was. Instead, the test solemnly announced that I better get used to being average.

Today, Thorndike's rank-obsessed educational labyrinth traps everyone within its walls—and not just the students. Teachers are evaluated at the end of each school year by administrators, and the resulting rankings are used to determine promotions, penalties, and tenure. Schools and universities are themselves ranked by various publications, such as *U.S. News and World Report,* who give great weight to the average test scores and GPA of the students, and these rankings determine where potential students will apply and what they're willing to pay. Businesses base their hiring decisions on applicants' grades and the ranking of their alma mater; these businesses are themselves sometimes ranked based on how many of their employees have advanced degrees and attended famous colleges. The educational systems of entire countries are ranked based on their national performance on international standardized tests such as the PISA (Programme for International Student Assessment) exam.[40]

Our twenty-first-century educational system operates exactly as Thorndike intended: from our earliest grades, we are sorted according to how we perform on a standardized educational curriculum designed for the average student, with rewards and opportunities doled out to those who exceed the average, and constraints and condescension heaped upon those who lag behind. Contemporary pundits, politicians, and activists continually suggest that our educational system is broken, when in reality the opposite is true. Over the past century, we have perfected our educational system so that it runs like a well-oiled Taylorist machine, squeezing out every possible drop of efficiency in the service of the goal its architecture was originally designed to fulfill: efficiently ranking students in order to assign them to their proper place in society.

A WORLD OF TYPE AND RANK

In the span of roughly fifty years—from the 1890s to the 1940s—virtually all our social institutions came to assess each of us in terms of our relationship to the average. During this transformative stretch, businesses, schools, and the government all gradually adopted the guiding conviction that the system was more important than the individual, offering opportunities to each of us according to our type or rank. Today, the Age of Average continues unabated. In the second decade of the twenty-first century, we are each evaluated according to how closely we approximate the average—or how far we are able to exceed it.

I'm not going to pretend that the Taylorization of our work-place and the implementation of standardization and rankings in our schools was some kind of disaster. It wasn't. When society embraced averagarianism, businesses prospered and consumers got more affordable products. Taylorism increased wages across society as a whole and probably lifted more people out of poverty than any other single economic development in the past century. By forcing college applicants and job seekers to take standardized tests, nepotism and cronyism were reduced and students from less privileged backgrounds attained unprecedented access to opportunities to a better life. Though it is easy to disparage Thorndike's elitist belief that society should divert resources toward superior students and away from inferior ones, he also believed that wealth and inherited privilege should play no part in determining a student's opportunities (on the other hand, he attributed different levels of mental talent to different ethnicities). Thorndike helped establish a classroom environment that made Americans out of millions of immigrants and raised the number of Americans with a high school diploma from 6 percent to 81 percent.[41] Overall, the universal implementation of averagarian systems across American society undoubtedly contributed to a relatively stable and prosperous democracy.

Yet, averagarianism did cost us something. Just like the Norma Look-Alike competition, society compels each of us to conform to certain narrow expectations in order to succeed in school, our career, and in life. We all strive to be like everyone else—or, even more accurately, we all strive to *be like everyone else, only better*. Gifted students are designated as gifted because they took the same standardized tests as everyone else, but performed better. Top job candidates are desirable because they have the same kinds of credentials as everyone else, only better. We have lost the dignity of our individuality. Our uniqueness has become a burden, an obstacle, or a regrettable distraction on the road to success.

We live in a world where businesses, schools, and politicians all insist that individuals really do matter, when everything is quite clearly set up so the system always matters more than you. Employees work for companies where they feel like they are being treated like cogs in the machine. Students get test results or grades that make them feel as if they will never attain their dreams. In our jobs and in school we are told there is one right way to get things done, and if we pursue an alternate course, we are often told that we are misguided, naive, or just plain wrong. Excellence, too often, is not prioritized over conforming to the system.

Yet we want to be recognized for our individuality. We want to live in a society where we can truly be ourselves—where we can learn, develop, and pursue opportunities on our own terms according to our own nature, instead of needing to conform ourselves to an artificial norm.[42] This desire prompts the billion-dollar question that drives this book: How can a society predicated on the conviction that individuals can only be evaluated in reference to the average ever create the conditions for understanding and harnessing individuality?

CHAPTER 3

OVERTHROWING THE AVERAGE

Peter Molenaar spent the early part of his long and acclaimed career as an averagarian scientist, building an international reputation through research on psychological development that relied heavily on average-based norms. He was so confident of the value of averagarian thinking that, on occasion, Molenaar needled colleagues who suggested that behavioral scientists might be relying too heavily on averages to understand individuals.[1]

Molenaar's confidence seemed justified. After all, he had spent a lifetime immersed in mathematics. In high school, he had been selected to compete in the Dutch Mathematical Olympiad. His dissertation for his doctorate in developmental psychology consisted of a mathematical tour de force describing "a dynamic factors model derived from a constrained spectral decomposition of the lagged covariance function into a singular and a nonsingular component." Molenaar's subsequent psychology publications were often so crammed full of equations and proofs that a lay reader might wonder if there was any psychology to be found at all.[2]

He rode his math talent and commitment to averagarianism to the very pinnacle of academic success in the Netherlands. In 2003, Molenaar was an H1 professor, the highest possible rank in the Dutch educational system, and served as the chairman of the Department of Psychological Methods at the prestigious University of Amsterdam. But in the Netherlands, the academic pinnacle has a shelf life. Dutch law mandates that all H1 professors must step back from their duties at the age of sixty-two to make way for their replacements, and then retire completely at age sixty-five. In 2003, Molenaar was fifty-nine, and he wasn't quite sure if he was ready to step down, but at least he was expecting to ride off into the academic sunset at the top of his game—until an unexpected request was laid at his feet.

With just three years to go as an active professor, he was somewhat annoyingly asked to take over a fall semester course after a colleague was abruptly sidelined. The class was a seminar on the theory and methods of mental testing. Trust me, the course is just as boring as it sounds. Much of test theory was codified into its modern form in a 1968 textbook often regarded as "the Bible of Testing," *Statistical Theories of Mental Test Scores,* authored by two psychometricians named Frederic Lord and Melvin Novick.[3] To this day it remains required reading for anybody who wants to design, administer, or understand standardized tests; I had to read Lord and Novick myself in grad school. It is the kind of book you gloss over as quickly as you can because the text is about as interesting as the instructions to a tax form. The tome is so dull, in fact, that nobody had ever noticed that its sleep-inducing pages concealed the thread that would unravel averagarianism.

To prepare for his stint as a substitute teacher, Molenaar opened his copy of Lord and Novick. That's when he experienced what he calls his *aha-erlebnis,* the German word for "epiphany." It was a moment that would change the direction of his life—and shake the foundations of the social sciences. In the book's introduction, Lord and Novick observed, rather drily, that any mental test attempted to discern the test taker's "true score" on some feature of interest. This made sense; isn't the reason we give someone an intelligence test, personality test, or college admissions test because we want to know their true intelligence ranking, true personality type, or true aptitude percentile?

Next, Lord and Novick observed that according to the leading theory of testing at the time—classical test theory[4]—the only way of determining a person's true score was by administering the *same* test to the *same* person, over and over again, many times.[5] The reason it was necessary to test someone repeatedly on, say, a math aptitude test, was because it was presumed that some amount of error would always occur in each test session (maybe the test taker was distracted or hungry; maybe she misread a question or two; maybe she guessed well). But if you took a person's average score across a large number of test sessions, this average score would converge onto the individual's true score.

The problem, as Lord and Novick fully acknowledged, was that it was impossible as a practical matter to test the same person multiple times, because human beings learn and therefore anyone who took, say, a math test, would inevitably perform differently as a result of seeing the same test again, foiling the possibility of obtaining many independent test scores.[6] But rather than admit defeat, Lord and Novick proposed an alternate way to derive someone's true score: instead of testing one person many times, they recommended testing many people one time.[7] According to classical test theory, it was valid to substitute a group distribution of scores for an individual's distribution of scores.

Adolphe Quetelet performed the same conceptual maneuver almost a century ago when he used the "Statue of the Gladiator" metaphor to define the meaning of a human average for the first time. He declared that taking the average size of 1,000 copies of a single soldier statue was equivalent to taking the average size of 1,000 different living soldiers. In essence, both Quetelet and Lord and Novick assumed that measuring one person many times was interchangeable with measuring many people one time.

This was Molenaar's moment of aha-erlebnis. He instantly recognized that Lord and Novick's peculiar assumption did not merely impact testing—the same assumption served as the basis for research in every field of science that studies individuals. It called into question the validity of an immense range of supposedly sturdy scientific tools: admissions tests for private schools and colleges; selection processes for gifted programs and special needs programs; diagnostic tests evaluating physical health, mental health, and disease risks; brain models; weight gain models; domestic violence models; voting behavior models; depression treatments; insulin administration for diabetics; hiring policies and employee evaluation, salary, and promotion policies; and basic methods of grading in schools and universities.

The strange assumption that a group's distribution of measurements could safely be substituted for an individual's distribution of measurements was implicitly accepted by almost every scientist who studied individuals, though most of the time they were hardly conscious of it. But after a lifetime of mathematical psychology, when Molenaar unexpectedly saw this unjustifiable assumption spelled out in black and white, he knew exactly what he was looking at: an irrefutable error at the very heart of averagarianism.

THE ERGODIC BAIT AND SWITCH

Molenaar recognized that the fatal flaw of averagarianism was its paradoxical assumption that you could understand individuals by ignoring their individuality. He gave a name to this error: "the ergodic switch." The term is drawn from a branch of mathematics that grew out of the very first scientific debate about the relationship between groups and individuals, a field known as ergodic theory.[8] If we wish to understand exactly why our schools, businesses, and human sciences have all fallen prey to a misguided way of thinking, then we must learn a little about how the ergodic switch works.

In the late 1800s, physicists were studying the behavior of gases. At the time, physicists could measure the collective qualities of gas molecules, such as the volume, pressure, and temperature of a canister of gas, but they had no idea what an individual gas molecule looked like or how it behaved. They wanted to know if they could use the average behavior of a group of gas molecules to predict the average behavior of a single gas molecule. To answer this question, physicists worked out a set of mathematical principles known as ergodic theory that specified exactly when you could use information about a group to draw conclusions about individual members of the group.[9]

The rules are fairly straightforward. According to ergodic theory, you are allowed to use a group average to make predictions about individuals if two conditions are true: (1) every member of the group is identical, and (2) every member of the group will remain the same in the future.[10] If a particular group of entities fulfills these two conditions, the group is considered to be "ergodic," in which case it is fine to use the average behavior of the group to make predictions about an individual. Unfortunately for nineteenth-century physicists, it turned out that the majority of gas molecules, despite their apparent simplicity, are not actually ergodic.[11]

Of course, you don't need to be a scientist to see that people are not ergodic, either. "Using a group average to evaluate individuals would only be valid if human beings were frozen clones, identical and unchanging," Molenaar explained to me.[12] "But, obviously, human beings are not frozen clones." Yet, even the most basic averagarian methods like ranking and typing all assumed that people *were* frozen clones. This was why Molenaar called this assumption the ergodic switch: it takes something nonergodic and pretends it is ergodic. We might think of the ergodic switch as a kind of intellectual "bait and switch" where the lure of averagarianism dupes scientists, educators, business leaders, hiring managers, and physicians into believing that they are learning something meaningful about an individual by comparing her to an average, when they are really ignoring everything important about her.

An example might help illustrate the practical consequences of the ergodic switch. Imagine you want to reduce the number of errors you make when you are typing on a keyboard by changing the speed at which you type. The averagarian approach to this problem would be to evaluate the typing skills of many different people, then compare the average typing speed to the average number of errors. If you do this, you will find that faster typing speeds are associated with fewer errors, on average. This is where the ergodic switch comes in: an averagarian would conclude that if *you* wanted to reduce the number of errors in your typing, then *you* should type faster. In reality, people who type faster tend to be more proficient at typing in general, and therefore make fewer errors. But this is a "group level" conclusion. If you instead model the relationship between speed and errors at the level of the individual—for instance, by measuring how many errors *you* make when typing at different speeds—then you will find that typing faster actually leads to *more* errors. When you perform the ergodic switch—substituting knowledge about the group for knowledge about the individual—you get the exact wrong answer.

Molenaar's epiphany also reveals the original sin of averagarianism, the mistake that occurred at the founding moment of the Age of Average: Quetelet's interpretation of the average size of Scottish soldiers. When Quetelet declared that his measurement of the average chest circumference actually represented the chest size of the "true" Scottish soldier, justifying this interpretation using the "Statue of the Gladiator," he performed the very first ergodic switch. This ergodic switch led him to believe in the existence of the Average Man and, even more important, was used to justify his assumption that *the average represents the ideal, and the individual represents error.*

A century and a half of applied science has been predicated on Quetelet's primal misconception.[13] That's how we ended up with a statue of Norma that matches no woman's body, brain models that match no person's brain, standardized medical therapies that target nobody's physiology, financial credit policies that penalize creditworthy individuals, college admission strategies that filter out promising students, and hiring policies that overlook exceptional talent.

In 2004, Peter Molenaar spelled out the consequences of the ergodic switch for the study of individuals in a paper entitled, "A Manifesto on Psychology as Idiographic Science: Bringing the Person Back into Scientific Psychology, This Time Forever."[14] After a scientific career devoted to averagarian thinking, his manifesto now declared that averagarianism was irredeemably wrong.

"I guess you could say I was like the biblical Paul," Molenaar told me with a smile. "At first, I was persecuting the Christians, all my colleagues who declared the average was wrong and the individual was the way. Then I had my 'road to Damascus' moment, and now I'm the biggest proselytizer of them all when it comes to the gospel of the individual."

THE SCIENCE OF THE INDIVIDUAL

Just because you're bringing the gospel to the gentiles doesn't mean they're going to listen. When I asked Molenaar about the initial reaction to his ideas, he replied, "As with most attempts to supplant or even slightly alter the ordained approaches, the arguments often fall on somewhat deaf ears. The more radical efforts can take on a Sisyphean character."[15]

Shortly after publishing his individuality manifesto, Molenaar gave a talk at a university about its details that included a call to move past averagarianism. One psychologist shook his head in response and declared, "What you are proposing is anarchy!"[16] This sentiment was perhaps the most common reaction among psychometricians and social scientists whenever Molenaar showcased the irreconcilable error at the heart of averagarianism. Nobody disputed Molenaar's math. In truth, it's fair to say that many of the scientists and educators whose professional lives were affected by the ergodic switch did not follow all the details of ergodic theory. But even those who understood the math and recognized that Molenaar's conclusions were sound still expressed the same shared concern: If you could not use averages to evaluate, model, and select individuals, well then . . . what *could* you use?

This practical retort underscores the reason that averagarianism has endured for so long and become so deeply ingrained throughout society, and why it has been so eagerly embraced by businesses, universities, governments, and militaries: because averagarianism worked better than anything else that was available. After all, types, ranks, and average-based norms are very convenient. It takes little effort to say things like "She is smarter than average," or "He was ranked second in his graduating class," or "She is an introvert," concise statements that seem true because they appear to be based on forthright mathematics. That is why averagarianism was a perfect philosophy for the industrial age, an era when managers—whether in businesses or schools—needed an efficient way to sift through large numbers of people and put them in their proper slots in a standardized, stratified system. Averages provide a stable, transparent, and streamlined process for making decisions quickly, and even if university administrators and human resource executives paid lip service to the problems associated with ranking students and employees, no manager was ever going to lose her job because she compared an individual to an average.

After hearing his colleagues react to his individuality manifesto by asking, quite reasonably, what they were supposed to use if they could not use averages, Molenaar realized it was not enough to prove averagarianism was wrong through some complex mathematical proof. If he truly wanted to overthrow the tyranny of the average once and for all, he needed to offer an *alternative* to averagarianism—some practical way to understand individuals that provided better results than ranking or typing.

Molenaar sat down with his boss, the dean of the graduate school at the University of Amsterdam, and excitedly informed her of his plans to develop a new scientific framework for studying and assessing individuals. He laid out ideas for several new projects, including an international conference on individuality, then asked for funding for these initiatives.

"You know I can't give you any new resources," the dean reluctantly replied. "You're stepping down in three years. I'm very sorry, Peter, you know the rules of the system, and there's nothing I can do."[17]

Molenaar was abruptly forced to look at himself in the mirror. Here he was, at the age of sixty, convinced he could make a profound contribution to science, one that could very possibly change the basic fabric of society. But launching revolutions was a young man's game, and the Dutch university system wasn't going to provide him with any support for his grand aspirations. He asked himself, Did he really want to fight for this?

Molenaar considered bowing to the inevitable—after all, he was on the tail end of a very successful career, and even if he did decide to become the leader of a game-changing scientific movement, it would demand not only years of research, but countless battles with scientists and institutions. But he didn't consider this too long. "When you realize what's at stake—how many parts of society are affected by all this," Molenaar told me, "I just had to try to find a way."[18]

He began seeking out new opportunities outside of the University of Amsterdam that would enable him to pursue his vision of building an alternative to averagarianism. In 2005, one materialized. On the other side of the Atlantic, Penn State University offered him a tenured position on their faculty, and shortly after that, appointed him the founding director of the Quantitative Developmental Systems Methodology core unit of the Social Science Research Institute, an entire research group he could shape as he wished. At Penn State, he gathered around himself a group of top-flight scientists and graduate students from around the world who shared his vision and who soon came to affectionately refer to Molenaar as "Maestro." Together, they began to lay the foundation for an actionable alternative to averagarianism: an interdisciplinary science of the individual.

Recall that the two defining assumptions of the Age of Average are Quetelet's conviction that *the average is the ideal, and the individual is error,* and Galton's conviction that *if someone is Eminent at one thing they are likely Eminent at most things.* In contrast, the main assumption of the science of the individual is that *individuality matters*[19]—the individual is not error, and on the human qualities that matter most (like talent, intelligence, personality, and character) individuals cannot be reduced to a single score.

Building on this new assumption, Molenaar and his colleagues began to develop new tools that enable scientists, physicians, educators, and businesses to improve the way they evaluate individuals. These tools often draw upon a very different kind of math than that used by averagarians. The mathematics of averagarianism is known as *statistics* because it is a math of *static values*—unchanging, stable, fixed values. But Molenaar and his colleagues argue that to accurately understand individuals one should turn to a very different kind of math known as *dynamic systems*—the math of changing, nonlinear, dynamic values.[20]

Since the assumptions and mathematics of the science of the individual are so different than those of averagarianism, it should be no surprise that the science of the individual also turns the method of studying people on its head.

ANALYZE, THEN AGGREGATE

The primary research method of averagarianism is *aggregate, then analyze:* First, combine many people together and look for patterns in the group. Then, use these group patterns (such as averages and other statistics) to analyze and model individuals.[21] The science of the individual instead instructs scientists to *analyze, then aggregate:* First, look for patterns within each individual. Then, look for ways to combine these individual patterns into collective insight. One example from developmental psychology illustrates how switching to an "individual first" approach to studying people can overturn long-standing convictions about human nature.

From the 1930s through the 1980s, scientists who studied infant development wrestled with a puzzling mystery known as the stepping reflex. When a newborn is held upright, she begins moving her legs in an up-and-down motion that closely resembles walking. For a long time, scientists suggested this stepping reflex pointed to the presence of an inborn walking instinct. But the reason this reflex was so mystifying was that at around two months of age, the reflex disappeared. When you hold up an older baby, her legs remain mostly motionless. But then, shortly before the infant begins to walk, the stepping reflex magically returns. What causes this reflex to appear, disappear, then appear again?

Scientists first attempted to solve the mystery of the stepping reflex using the traditional method of averagarianism: aggregate, then analyze. Since everybody presumed that the stepping reflex was associated with neural development, scientists examined a large number of infants, calculated the average age that the stepping reflex appeared and disappeared, then compared these average ages with the average age of various milestones of neural development. Scientists discovered that one neural process seemed to correspond with the appearance and disappearance of the stepping reflex: myelination, the physiological process whereby neurons grow a protective sheathing. So scientists proposed the "myelination theory": each baby is born with the stepping reflex, but as the motor control center of the brain begins to myelinate, the reflex vanishes. Then, after the motor control center of the brain develops further, the baby regains conscious control of the reflex.[22]

By the early 1960s, myelination theory had become the standard medical explanation of the stepping reflex. It even served as the basis for the diagnosis of neural disorders: if a baby's stepping reflex did not disappear on time, physicians and neurologists warned the parents that their child might have some kind of neurological disability.[23] Many pediatricians and child psychologists even asserted it was bad for parents to try to encourage their child's stepping reflex, arguing it could delay normal development and cause neuromuscular abnormalities.

The curious and unwieldy myelination theory held sway over American pediatrics for several decades and might have even lasted into the twenty-first century, if not for a young scientist named Ester Thelen.[24] While studying animals early in her career, Thelen discovered that many instinctive behaviors that biologists insisted were fixed and rigid were actually highly variable, depending in large part on the unique quirks of each individual animal. These formative professional experiences drove her to study the mathematics of dynamic systems, and eventually she decided to reexamine the human stepping reflex by focusing on the individuality of each child.

Thelen studied forty babies over a period of two years. Every day she took a photo of each baby, examining their individual physical development. She held them over treadmills and placed them in different positions to analyze the individual mechanics of each baby's motions. Eventually, she formulated a new hypothesis about what was causing the disappearance of the stepping reflex: chubby thighs.

She noticed that the babies who gained weight at the slowest rate tended to move their legs the most and for the longest period of time. The babies who gained weight at the fastest rate tended to lose their stepping reflex the earliest, because their leg muscles were simply not strong enough to lift up their legs. It was not the *absolute* chubbiness of the thighs that was the key factor, but rather the *rate* of physical growth, since what mattered was the amount of body fat relative to the development of muscle strength.[25] That is why previous scientists who had simply compared average ages to average weights had never discovered anything. The aggregate, then analyze approach disguised each child's individual pattern of development. Thelen's analyze, then aggregate approach revealed it.

Needless to say, chubby thighs had never before appeared in any scientific account of the stepping reflex, so many researchers rejected the idea out of hand. But in a series of ingenious experiments, Thelen proved beyond a doubt that her theory of chubby thighs was correct. She placed babies in water and voilà—the stepping reflex reemerged, even in those babies with the fattest legs. She added different weights to babies' legs and was able to predict which babies would lose the stepping reflex.[26]

When Esther Thelen studied the individuality of each baby, she arrived at an explanation that had eluded generations of averagarian researchers who had informed parents that there might be something wrong with their infant's brain when the real cause of their concerns was their infant's flabby thighs.

At Penn State, Peter Molenaar and his department have demonstrated a number of similar findings where an individual-first approach leads to superior results compared to relying only on group averages. There is one difficulty presented by this individual-first approach: it requires a great deal of data, far more data than averagarian approaches. In most fields that study human beings, we didn't have the tools one hundred, fifty, or even twenty-five years ago to acquire and manage the extensive data necessary to effectively analyze, then aggregate. During the industrial age, averagarian methods were state of the art, and individual-first methods were often mere fantasy. But now we live in the digital age, and over the past decade the ability to acquire, store, and manipulate massive amounts of individual data has become convenient and commonplace.

All that's missing is the mindset to use it.

INDIVIDUALITY MATTERS

When Lieutenant Gilbert Daniels first suggested that cockpits needed to fit *every* pilot instead of the *average* pilot, it seemed an impossible task. Today, the same companies that once said it could not be done tout the flexibility of their cockpits as a selling point.[27] Similarly, when Esther Thelen decided to challenge the deeply entrenched myelination theory by studying the individuality of babies, it seemed a difficult undertaking at the very least, and possibly a pointless one. But it did not take long for her to notice the overlooked role of chubby thighs.

Averagarianism forces our thinking into incredibly limiting patterns—patterns that we are largely unaware of, because the opinions we arrive at seem to be so self-evident and rational. We live in a world that encourages—no, *demands*—that we measure ourselves against a horde of averages and supplies us with no end of justification for doing so. We should compare our salary to the average salary to judge our professional success. We should compare our GPA to the average GPA to judge our academic success. We should compare our own age to the average age that people get married to judge whether we are marrying too late, or too early. But once you free yourself from averagarian thinking, what previously seemed impossible will start to become intuitive, and then obvious.

It's easy to have sympathy for the psychologist who informed Molenaar, "What you are proposing is anarchy!" Letting go of the average seems unnatural. It's venturing beyond the known shores, a proposition that seems particularly foolhardy when the entire world around you remains firmly on the terra firma of averagarianism. But there's no need to grope blindly in the dark. In the next part of the book, I will share three principles drawn from the science of the individual that can replace your reliance on the average: the jaggedness principle, the context principle, and the pathways principle. These principles will allow you to evaluate, select, and understand individuals in a whole new way. They will allow you to discard types and ranks and discover instead the true patterns of individuality in your own life. They will help you eliminate the unchallenged authority of the average once and for all.

PART II

THE PRINCIPLES
OF INDIVIDUALITY

*An individual is a high-dimensional system
evolving over place and time.*

—Peter Molenaar, Pennsylvania State University

CHAPTER 4

TALENT IS ALWAYS JAGGED

By the mid–2000s, Google was already well on its way to becoming an era-defining Internet juggernaut and one of the most innovative and successful corporations in history. To sustain its extraordinary levels of growth and innovation, Google had a voracious appetite for talented employees. Fortunately, the company was flush with cash, and the combination of high salaries, generous perks, and the chance to work on innovative products made Google one of the world's most desirable places to work.[1] By 2007, it was receiving a hundred thousand job applications every month, ensuring that Google would have its choice of top talent—as long as it could figure out how to *identify* that top talent.[2]

At first, Google made its hiring decisions the same way that most Fortune 500 companies did: by looking at each job applicant's SAT scores, academic GPA, and diploma and hiring those applicants who ranked at the very top.[3] Before long, the Google campus at Mountain View was full of employees with near-flawless SATs, valedictorian-level grades, and advanced degrees from the likes of Caltech, Stanford, MIT, and Harvard.[4]

Ranking individuals on a handful of metrics—or even a single metric—is not only common practice when recruiting new employees, it is the most prevalent method of evaluating existing employees, too.[5] In 2012, Deloitte, the largest professional services firm in the world, assigned every one of its more than sixty thousand employees a numerical rating based on their performance on work projects, and then at the end of the year at a "consensus meeting" these project ratings were combined into a final rank ranging from 1 to 5. Each employee, in other words, was evaluated using a single number. It is hard to imagine a more straightforward method for comparing employees' value than assessing them on a simple, one-dimensional scale.[6]

According to the *Wall Street Journal,* an estimated 60 percent of Fortune 500 firms still used some form of single-score ranking systems to evaluate employees in 2012.[7] Perhaps the most extreme version of these systems is what is known as "forced ranking," a method pioneered by General Electric in the 1980s where it was known as "rank and yank."[8] In a forced ranking system, employees are not merely ranked on a one-dimensional scale; a certain predetermined percentage of employees must be designated as above average, a certain percentage must be designated as average—and a certain percentage *must* be designated as below average. Those employees assigned to the top ranks receive bonuses and promotions. Those at the bottom receive warnings or, in some cases, are simply let go.[9] By 2009, 42 percent of large companies were using forced ranking systems, including Microsoft, whose well-publicized version was known as "stack ranking."[10]

Of course, it's easy to understand why so many businesses have adopted single-score systems for hiring and performance evaluation: they are easy and intuitive to use and they carry the imprint of objectiveness and mathematical certitude. If an applicant is ranked higher than average, then hire her or reward her. If she ranks lower, pass on her or let her go. If you want more talented employees, simply "raise the bar"—increase the score you use as your cutoff for hiring or promotions.

Ranking individual talent and performance on a single scale, or a few scales, seems to make perfect sense. And yet, by 2015, Google, Deloitte, and Microsoft had each modified or abandoned their rank-based hiring and evaluation systems.

Despite Google's continued growth and profitability, by the mid-2000s there were signs that something was wrong with the way it was selecting talent. Many of its hires were not performing the way management had imagined, and there was a growing sense within Google that company recruiters and managers were ignoring many candidates whose talent was not getting captured by the familiar metrics used by most companies, such as grades, test scores, and diplomas.[11] As Todd Carlisle, the human resources director for product quality operations at Google, explained to me, "We began to spend a lot of time and money analyzing the 'missed talent' that we felt we should have hired, but didn't."[12]

At Deloitte, by 2014 they were also beginning to realize that its single-score employee evaluation method was not working as well as they expected. Deloitte was devoting more than two million hours each year to the process of calculating employee performance rankings—a tremendous amount of time—but the value of these rankings was being questioned.[13] In a *Harvard Business Review* article coauthored with Marcus Buckingham, Ashley Goodall, former director of leader development at Deloitte, wrote that what gave them pause was research suggesting that a single-score rating might not capture the true performance of an employee so much as reveal the idiosyncratic tendencies of the person rating that performance.

"Both internally and externally it was clear that people were starting to recognize that traditional single-score performance reviews don't work, so there was a sense of clarity about what we needed to get away from," Goodall told me.[14]

Meanwhile, at Microsoft, stack ranking was an unmitigated disaster. A 2012 *Vanity Fair* article called the era when Microsoft relied on stack ranking "the lost decade." The performance rating system forced employees to compete for rankings, killing collaboration among employees and, worse, leading employees to avoid working with top performers, since doing so threatened to lower their own ranking as a result. While stack ranking was in effect, the article reports, the company had "mutated into something bloated and bureaucracy-laden, with an internal culture that unintentionally rewards managers who strangle innovative ideas that might threaten the established order of things."[15] In late 2013, Microsoft abruptly jettisoned stack ranking.[16] So where did Google, Deloitte, and Microsoft go wrong?

Each of these innovative companies initially followed the averagarian notion that you can effectively evaluate individuals by ranking them—a notion rooted in Francis Galton's belief that if you are good or eminent at one thing, you are good or eminent at most things.[17] And to most of us, it seems like this approach *ought* to have worked. After all, isn't it obvious that some people are generally more talented than others, and therefore it should be possible to rank talent on a single scale and make assumptions about their potential based on that ranking? Google, Deloitte, and Microsoft, however, discovered that the idea that talent can be boiled down to a number that we can compare to a neat average simply doesn't work. But why? What is at the root of the unexpected failure of ranking?

The answer is one-dimensional thinking. The first principle of individuality—the *jaggedness principle*—explains why.

THE JAGGEDNESS PRINCIPLE

Our minds have a natural tendency to use a one-dimensional scale to think about complex human traits, such as size, intelligence, character, or talent. If we are asked to assess a person's size, for example, we instinctively judge an individual as large, small, or an Average Joe. If we hear a man described as *big,* we imagine someone with big arms and big legs and a big body—someone who is large all over. If a woman is described as *smart,* we assume she is likely good at solving problems across a wide range of domains, and is probably well educated, too. During the Age of Average, our social institutions, particularly our businesses and schools, have reinforced our mind's natural predilection for one-dimensional thinking by encouraging us to compare people's merit on simple scales, such as grades, IQ scores, and salaries.[18]

But one-dimensional thinking fails when applied to just about any individual quality that actually matters—and the easiest way to

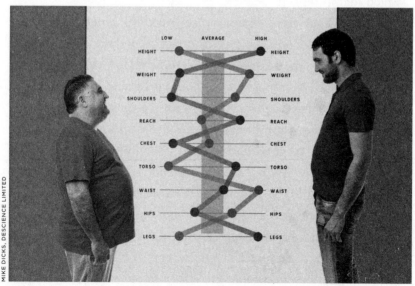

JAGGED PROFILES FOR SIZE

understand why is to take a closer look at the true nature of human size. The previous image portrays the measurements of two men on nine dimensions of physical size, the same dimensions analyzed by Gilbert Daniels in his breakthrough study of pilots.

Which man is bigger? It seems like there should be an easy answer, but when you compare the two men on each dimension, the answer turns out to be more elusive than we might expect. The man on the right is tall but has narrow shoulders. The man on the left has a large waist but nearly average-size hips. You might attempt to determine which man is bigger by simply taking the average of all nine of each man's dimensions—except if you performed this calculation, you would discover that each man's average size is nearly identical. At the same time, we can see that it would be misleading to say they are the same size—or to describe either one of them as average: the man on the left is average on two dimensions (reach and chest), while the man on the right is barely average on only one dimension (waist). There is no simple answer to the question, "Which man is bigger?"

While this may seem obvious once you think about it, don't let this statement fool you, because the fact that there is no answer to the question—the reason it is not possible to rank individuals on size—reveals an important truth about human beings and the first principle of individuality: the *jaggedness principle*. This principle holds that we cannot apply one-dimensional thinking to understand something that is complex and "jagged." What, precisely, is jaggedness? A quality is jagged if it meets two criteria. First, it must consist of multiple dimensions. Second, these dimensions must be weakly related to one another. Jaggedness is not just about human size; almost every human characteristic that we care about—including talent, intelligence, character, creativity, and so on—is jagged.

To understand these criteria, let's return to our example of human size. If the question we were trying to answer was "Which man is *taller?,"* the answer would be easy. Height is one-dimensional, so it is perfectly acceptable to rank people according to how tall they are. But human size is a different story: it is composed of many different dimensions that are not strongly linked. Look again at the figure. The vertical band in the center of the image represents the range of measurements of the "average pilot" as Daniels had once defined it. For decades, the Air Force presumed that the bodies of most pilots would lie within that vertical band, because they assumed that someone with average-size arms would also have average-size legs and an average-size torso. But, because size is jagged, it turns out this is not true at all. In fact, Daniels discovered that less than 2 percent of pilots measured were average on four or more of these nine dimensions, and nobody was average on all of them.[19]

What if we were to expand the average band to include the middle 90 percent of each dimension, instead of the middle 30 percent? You might guess that most people's bodies would surely lie within such a wide range. In actuality, less than half of all people would.[20] It turns out that *most* of us have at least one body part that is rather large or rather small. That is why a cockpit designed for the average is a cockpit designed for nobody. Jaggedness also explains why the Norma-Look-Alike contest organizers could not find a woman who was a perfect match. Women have long protested the artificially exaggerated dimensions of Mattel's Barbie doll, but the principle of jaggedness tells us that an average-size doll—a Norma-size doll—is just as phony.

Of course, it is reasonable to sometimes *pretend* size is one-dimensional if the trade-off is worth it, like when it comes to mass-produced clothing: in return for a lack of great fit for any one person, we get inexpensively manufactured shirts and pants for everyone. But if the stakes are high—if you're altering an expensive wedding gown or designing a safety feature like an automobile airbag, or engineering the cockpit of a jet—then ignoring the multidimensionality of size is never a good compromise. When it matters, there are no shortcuts: you can only produce a good fit if you think about size in terms of all its dimensions.

Just about any meaningful human characteristic—especially talent—consists of multiple dimensions. The problem is that when trying to measure talent, we frequently resort to the average, reducing our jagged talent to a single dimension like the score on a standardized test or grades or a job performance ranking. But when we succumb to this kind of one-dimensional thinking, we end up in deep trouble. Take, for example, the New York Knicks.

In 2003, Isiah Thomas, a former NBA star, took over as president of basketball operations for the Knicks with a clear vision of how he wanted to rebuild one of the world's most popular sports franchises. He evaluated players using a one-dimensional philosophy of basketball talent: he acquired and retained players based solely on the average number of points they scored per game.[21]

Thomas figured that since a team's basketball success was based on scoring more points than your opponent, if your players had the highest combined scoring average, you would expect—on average—to win more games. Thomas was not alone in his infatuation with top-ranked scoring. Even today a player's scoring average is usually the most important factor in determining salaries, postseason awards, and playing time.[22] But Thomas had made this single metric the most important factor for selecting *every* member of the team, and the Knicks had the financial resources to make his priority a reality. In effect, the Knicks were assembling a team using the same one-dimensional approach to talent that companies use when making academic rankings the primary criteria for hiring employees.

At great expense, the Knicks managed to assemble a team with the highest combined scoring average in the NBA . . . and then suffered through four straight losing seasons, losing 66 percent of their games.[23] These one-dimensional Knicks teams were so bad that only two teams had a worse record during the same stretch. The jaggedness principle makes it easy to see why they failed so badly: because basketball talent is multidimensional. One mathematical analysis of basketball performance suggests that at least five dimensions have a clear effect on the outcome of a game: scoring, rebounds, steals, assists, and blocks.[24] And most of these five dimensions are not strongly related to one another—players who are great at steals, for instance, are usually not so great at blocking. Indeed, it is exceptionally rare to find a true "five-tool player." Out of the tens of thousands of players who have come through the NBA since 1950 only five players have ever led their team on all five dimensions.[25]

The most successful basketball teams are composed of players with complementary profiles of basketball talent.[26] In contrast, Thomas's Knicks teams were terrible at defense and, perhaps surprisingly, they were not even particularly great at offense despite the talented scorers on the team, since each individual player was more intent on getting his own shots than facilitating anyone else's. The Knicks—like Google, Deloitte, and Microsoft—eventually realized that a one-dimensional approach to talent was not producing the results they wanted. After Thomas left in 2009, the Knicks returned to a multidimensional approach to evaluating talent and started winning again, culminating in a return to the playoffs in 2012.[27]

THE WEAKEST LINKS

For a human trait like size or talent to be considered jagged, however, it's not enough to be multidimensional. Each of the dimensions also must be relatively independent. The mathematical way to express this independence is *weak correlations*.

Francis Galton helped develop the statistical method of correlation more than a century ago as a way of assessing the strength of the relationship between two different dimensions, like height and weight.[28] Galton began applying an early version of correlation to people with the intention of demonstrating the validity of his idea of rank: that a person's talent, intelligence, health, and character were closely related to one another.[29] Today, we express correlation as a value between 0 and 1, where 1 is a perfect correlation (like the correlation between your height in inches and your height in centimeters) and 0 is no correlation at all (like the correlation between your height in inches and the temperature on Saturn).[30] Across many scientific fields, a correlation of 0.8 or higher is considered strong while a correlation of 0.4 or lower is considered weak, although the precise cutoffs for "strong" and "weak" are ultimately arbitrary.

If the correlations between all the dimensions in a system are strong, then that system is not jagged and you are justified in applying one-dimensional thinking to make sense of it. Consider the Dow Jones Industrial Index. The Dow is a single numerical score that represents the combined stock value of thirty large and famous "blue-chip" companies. At the close of each American business day, the financial news dutifully reports the value of the Dow to the hundredth decimal place (it was 17,832.99 on January 2, 2015) and whether this number has moved up or down. Investors use the Dow to evaluate the overall strength of the stock market, and for good reason—between 1986 and 2011 (twenty-five years), the average correlation between the Dow and four other leading stock market indices was extremely high: 0.94.[31] Even though the stock market is multidimensional (there are thousands of publicly traded companies in the United States), its general vitality can be reasonably captured with a single number: using the Dow to assess the overall strength of the stock market is one-dimensional thinking at its most reasonable.

Human size, however, is a different matter. In 1972, in a follow-up to Daniels's study of pilots, U.S. Navy researchers calculated the correlations between ninety-six dimensions of Naval aviators' size. They found that only a few correlations were stronger than 0.7, while many were lower than 0.1. The average correlation among all ninety-six dimensions of body size for Naval aviators was only 0.43.[32] This means that knowing someone's height or neck thickness or grip width is unlikely to tell you much about the rest of his dimensions. If you want to truly understand a person's body size, there is no simple way to summarize it. You need to know the details of their jagged profile.

What about our minds? Are mental abilities jagged? When Galton first introduced correlation into the social sciences, he did so with the expectation that scientists would find *strong* correlations between our mental abilities—that our minds, in other words, were not very jagged.[33] One of the very first scientists to systematically test this hypothesis was a man named James Cattell, the first American to obtain a Ph.D. in psychology and an early pioneer of testing theory who coined the term "mental test."[34] He was also an enthusiastic disciple of Galton's idea of rank. In the 1890s, Cattell set out to prove, once and for all, that a one-dimensional view of mental ability was justified.[35]

Cattell administered a battery of physical and mental tests to hundreds of incoming freshmen at Columbia University across several years, measuring such things as their reaction time to sound, their ability to name colors, their ability to judge when ten seconds passed, and the number of letters in a series they could recall. He was convinced he would discover strong correlations between these abilities—but, instead, he found the exact opposite. There was virtually no correlation at all.[36] Mental abilities were decidedly jagged.

For a devout believer in ranking, there was worse to come. Cattell also measured the correlations between students' grades in college courses and their performance on these mental tests and discovered very weak correlations between them. And not only that—even the correlations between students' grades in different classes were low. In fact, the only meaningful correlation Cattell found at all was between students' grades in Latin classes and their grades in Greek classes.[37]

At the dawn of our modern educational system, when our schools were first becoming standardized around the mission of sorting students into average, above-average, and below-average bins of "general talent," the first scientific investigation of this assumption revealed that it was false. But psychologists were so convinced that one-dimensional mental talent *must* exist, even if it was hidden, that most of Cattell's colleagues rejected his results, suggesting that something was wrong with the way he conducted his experiments or analyzed his results.[38]

Meanwhile, psychologists—and then education, and then business—all doubled down on the notion that mental abilities are highly correlated and could be represented with a one-dimensional value like an IQ score.[39] Ever since Cattell, study after study has revealed that individual intelligence—not to mention personality and character—is jagged.[40] Even Edward Thorndike, who fashioned our modern education system around the notion that if you are good at one thing, then you are good at most things, conducted his own research to examine the correlation between school grades, standardized test scores, and success at professional jobs. He also found weak correlations between all three—yet he still rationalized that he could safely ignore this fact because he believed in a hypothetical (though unproven) one-dimensional "learning ability" that undergirded success in both school and work.[41]

Even today, scientists, physicians, businesspeople, and educators rely on the one-dimensional notion of an IQ score to evaluate intelligence. Even if we're willing to concede that, yes, there are multiple kinds of intelligence—like musical intelligence, or artistic intelligence, or athletic intelligence—it's hard to shake the feeling that there must be some kind of "general intelligence" a person possesses that can be applied to a great many domains. If we hear that one person is smarter than another, we assume that the smarter person is probably going to do better at just about any intellectual task that we set before him or her.

Consider, though, the following two jagged profiles of intelligence. They show scores for two different women on the Wechsler Adult Intelligence Scale (WAIS),[42] one of the two most commonly used contemporary tests of intelligence.[43] Each woman's profile represents her score on ten subtests from the WAIS test, each measuring a different dimension of intelligence, such as vocabulary or puzzle solving. All of the subtest scores are combined to generate an individual's IQ score.

Which woman is smarter? According to the WAIS, they are equally intelligent—each has an IQ of 103—and each is close to average intelligence, defined as an IQ of 100. If we were tasked with hiring the smartest candidate for a job, we might rate each woman equally. Yet each of these women clearly possesses different mental strengths and weaknesses, and if the goal is to understand

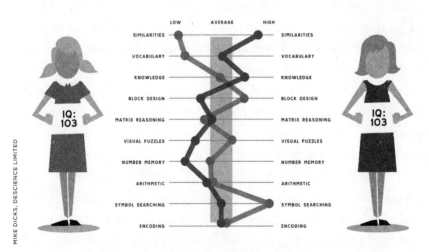

JAGGED PROFILES FOR INTELLIGENCE

these women's talents, it is obvious that relying on an IQ score is misleading.[44]

As with physical size, the correlations between each of the dimensions of mental ability assessed by the WAIS are for the most part not particularly strong,[45] indicating that mental talent is jagged and cannot be described or understood by a one-dimensional value like an IQ score. Yet to this day few of us can resist the lure of evaluating a person's intelligence with a single ranking or number. But a one-dimensional evaluation of mental abilities is even more misguided than these intelligence profiles portray. If you subdivide intelligence even further and compare, for instance, short-term memory for words to short-term memory for images, scientists have shown that these "microdimensions" also exhibit weak correlations.[46] No matter how fine you slice your mind, you are jagged all the way down.

All of this leads to one obvious question: If human abilities are jagged, why do so many psychologists, educators, and business executives continue to use one-dimensional thinking to evaluate talent? Because most of us have been trained in averagarian science, which implicitly prioritizes the system over the individual. It is entirely possible to build a functional evaluation system upon weak correlations: if you select employees based on a one-dimensional view of talent, while you may be wrong about any one individual, *on average* you will do better than someone who selects employees randomly.

As a result, we have managed to convince ourselves that weak correlations mean something that they do not. In most fields of psychology and education, if you find a correlation of, say, 0.4 (the correlation between SAT scores and first-semester college grades[47]), it is usually assumed you have found something important and meaningful. Yet, according to the mathematics of correlation, if you find a 0.4 correlation between two dimensions, that means you have managed to explain 16 percent of the behavior of each dimension.[48] Do you really understand something if you can explain 16 percent of it? Would you hire a mechanic who said he could explain 16 percent of what was wrong with your car?

Of course, if we care more about the efficiency of the system than about individuality, then understanding 16 percent, on average, is undeniably better than nothing. It may even be enough to set policy for *groups* of people. But if our goal is to identify and nurture individual excellence, then weak correlations tell us something different: we will only succeed if we pay attention to the distinct jaggedness of every individual.

OVERCOMING TALENT BLINDNESS

In 2004, Todd Carlisle became an analyst in the Google human resources department, where he helped facilitate interactions between Google project managers who needed to hire new employees and recruiters who put together "hiring packets" about job candidates that the managers could use to make their hiring decisions. At the time, a candidate's GPA and standardized test scores held a prominent place in these packets. But Carlisle noticed a very curious phenomenon: increasingly, project managers were asking recruiters to include additional information about the candidates.[49] Some wanted to know whether the candidates had competed in programming competitions. Others wanted to know if their hobbies included chess or playing in a band. It seemed every project manager had a pet idea about what extra information was salient when making a hiring decision.

"One day I just realized, if the traditional metrics—the grades and test scores—were really so great, why was everyone supplementing them with additional, clearly nontraditional metrics?" Carlisle told me. "That's when I decided to do the experiment."[50] Carlisle harbored the private feeling that there were probably a lot of talented people out there that Google was missing out on, and he thought part of the problem was an overemphasis on a small set of familiar metrics. He believed he could change the way the company approached recruiting so that it instead looked at the whole applicant in all her complexity. Since big decisions at Google operate primarily by consensus rather than decree, Carlisle knew that if he was going to convince project managers of the value of his multidimensional vision of talent evaluation, he would need a study that systematically tested not only his own ideas about which dimensions of talent predicted success at Google, but *all* the dimensions that managers and executives believed were related to being a great employee.

First, Carlisle collected an enormous list of more than three hundred dimensions (he called them "factors") that included traditional dimensions like standardized test scores, diplomas, alma mater rankings, and GPAs, as well as more idiosyncratic factors that other managers had identified as being significant. (One prominent Google executive, for example, suggested that the age someone first became interested in computers might be important.) Next, Carlisle ran test after test to analyze which of these factors was actually related to employee success. The results were startling and unequivocal.[51]

It turned out that SAT scores and the prestige of a candidate's alma mater were not predictive at all. Neither was winning programming competitions. Grades mattered a little, but only for the first three years after you graduated. "But the real surprise for me and for a lot of people at Google," Carlisle told me, "was that when we analyzed the data we couldn't find a single variable that mattered for even most of the jobs at Google. Not one."[52]

In other words, there were many different ways to be talented at Google, and if the company wanted to do the best possible job of recruiting employees, it needed to be sensitive to all of them. Carlisle had discovered the jaggedness of Google talent and, as a result, made changes to the way Google recruits new employees.

They rarely ask candidates for their GPAs if they've been out of school for three years and no longer require test scores for any candidate. "We no longer look at school selectivity the same way, either," Carlisle explained to me. "The challenge now is not only what information to collect, but how to present it—you have to focus on which factors you emphasize as most important in a hiring packet. The experiment has helped to create a more complete picture of candidates that managers can use to make better hires."[53]

Taking into consideration the jagged talent of job applicants is not some kind of sophisticated luxury that only giants like Google can afford to undertake. It's also a way for smaller companies to identify and attract top talent in a competitive job market. IGN is a popular website devoted to video games and other media, but it has less than 1 percent the number of employees of Google, and an even smaller percentage of sales.[54] Initially, IGN approached hiring using the same one-dimensional thinking as other tech companies. Of course, if every company in the entire tech industry is evaluating employees using identical one-dimensional criteria like grades and standardized test scores, there is going to be a very small set of candidates at the top of these rankings—and these "top-ranked" candidates are far more likely to sign with a big fish like Google or Microsoft than a small one like IGN.

IGN executives realized they simply couldn't compete with all the other tech firms for the employees they considered talented. They only had two choices: offer higher pay—not feasible—or change the way they thought about talent. So in 2011, IGN created a program called Code-Foo that was a "no résumé allowed" recruitment program aimed at finding untapped programming talent.[55] The six-week program paid aspiring programmers to learn new programming languages and then work on actual software engineering projects at IGN.[56] What was so unusual about Code-Foo was the way IGN managers evaluated applicants. They completely ignored applicants' educational background and previous experience. Instead of submitting a résumé, candidates submitted a statement of passion for IGN and answered four questions that tested their coding ability. In essence, IGN was saying, "We don't care what you've done or how you learned to program, we just want you to be good—and excited about putting your skills to work."

In 2011, 104 people applied to the Code-Foo program; 28 were accepted, and only half of them had earned a college degree in a technical field. IGN president Roy Bahat told *Fast Company* magazine that he hoped Code-Foo would eventually lead to one or two hires. IGN ended up hiring eight.[57] "It's not like if you looked at their résumés, you would have said it's impossible that they would be qualified for the jobs," Bahat reported to *Fast Company*. "But if you only looked at their résumés . . . there wouldn't necessarily be a reason to say yes. They're the kind of people we would have overlooked."[58]

Often, when organizations embrace jaggedness for the first time, they feel like they have found a way to uncover diamonds in the rough, to identify unorthodox or hidden talent. But the jaggedness principle says otherwise: while we may have identified *overlooked* talent, there is nothing unorthodox or hidden about it. It is simply true talent, as it has always existed, as it can only exist in jagged human beings. The real difficulty is not finding new ways to distinguish talent—it is getting rid of the one-dimensional blinders that prevented us from seeing it all along.

Of course, the blinders that are most important to eliminate are the ones we use to look at ourselves.

TAPPING INTO YOUR FULL POTENTIAL

As I approached the end of my degree requirements at Weber State University, I decided to apply to graduate schools in fields related to neuroscience. If I could get admitted, I would become the first person on either side of my family to attend graduate school. I had managed to turn things around in college and obtain strong grades as well as enthusiastic letters of recommendation from a couple of professors. Only one thing was standing in my way: a standardized test.

I needed to perform well on the GRE, the Graduate Record Examination, a test required by every one of the graduate science programs I was applying to.[59] At the time, the test consisted of three parts: a math part, a verbal part, and the so-called analytical reasoning part, which is supposedly designed to evaluate your ability to think logically. It consisted of knotty word problems along the lines of, "Jack, Jenny, Jeanie, Julie, Jerry, and Jeremy are all attending a dinner party. Jack doesn't like Jenny, Jeanie doesn't like Jeremy, Julie loves Jerry, and Jenny always steals Julie's dinner rolls. If they are sitting at a round table, who would you seat to Jeremy's left?"

I started preparing for the GRE six months before I needed to take the exam, but with just two weeks to go things were looking grim. I had taken about twenty practice tests. I consistently did well on the math and verbal sections, but the analytical reasoning section was a disaster. I never scored above the 10th percentile. Each time, I got almost every question wrong. My tutor, who got a perfect score on the analytical reasoning section, had shared his method with me, and I figured that if I simply practiced his method enough times, eventually my performance would improve. It didn't. The Julies and Jennys and Jeanies all blurred together and I could never seem to reason my way through to the answer. I was once again staring at the possibility of all my dreams coming to an unceremonious end, because it was hard to imagine any graduate program would admit someone who scored in the 10th percentile on any test.

While studying at my parent's house, I got so frustrated that I chucked my pencil across the room, nearly spearing my father as he unexpectedly walked by. Lucky for me, he came over and asked what was going on. I told him I was failing the analytical section and showed him the method I was using to solve the problems.

"That requires you to do most of the problem in your head," he pointed out.

"Sure," I responded. "That's the way the problems are supposed to be done." *After all*, I thought to myself, *my teacher had got a perfect score with that method, and most of the other kids in my test preparation class were also scoring above the 80th percentile using it.*

"But you don't have great working memory. Why would you try to use a method that places demands on your working memory?" he said. He knew, however, that I had done well in my geometry classes. "You're pretty good at visual thinking, though, so why don't you use a problem-solving method that relies on that."

He sat down and proceeded to show me a way to convert each problem into a kind of visual table that allowed me to draw the precise relationships between Jerry and Jenny and Julie in a clear and reliable fashion. At first, I was completely skeptical that this technique, which was indeed very easy for me, could possibly work. But I tried it out on problem after problem, and each time it gave me the right answer. I couldn't believe it. Two weeks later I took the GRE and I got my highest score on the analytical reasoning section.

My GRE instructor had figured out a way to solve problems that suited his jagged mental abilities—but not necessarily mine. Fortunately, my father had a clearer sense of my jaggedness. He helped me see that my problem was not that I had weak analytical skills—the one-dimensional view I had settled on after failing practice test after practice test using my instructor's method—but rather that I was relying on one of my weakest mental abilities, working memory, to solve the problems. Once my father helped me identify a strategy that played to my strengths, I could finally answer the test questions correctly and demonstrate my true talent.

I owe a debt of gratitude to my dad. His thoughtful consideration of my jagged profile—my individuality—led him to offer invaluable advice that changed the course of my life. If I had not switched to a visual way of analyzing the GRE problems, I would have performed poorly on the test and, as a result, would probably never have gotten into Harvard. That is the power of the first principle of individuality. When we are able to appreciate the jaggedness of other people's talents—the jagged profile of our children, our employees, our students—we are more likely to recognize their untapped potential, to show them how to use their strengths, and to identify and help them improve their weaknesses, just like my dad did.

And when we become aware of our own jaggedness, we are less likely to fall prey to one-dimensional views of talent that limit what we are capable of. Had I failed the test, it's likely that I would have concluded that I didn't have what it takes to succeed in graduate school—after all, that's what the test is *supposed* to tell you—and lowered the expectations that I had for myself.

Recognizing our own jaggedness is the first step to understanding our full potential and refusing to be caged in by arbitrary, average-based pronouncements of who we are expected to be.

CHAPTER 5

TRAITS ARE A MYTH

A re you an extrovert or an introvert? This deceptively simple question plunges us into one of the oldest and most contentious debates in psychology: the nature of your personality. On one side of the debate are the *trait psychologists,* who argue that our behavior is determined by well-defined personality traits, such as introversion and extroversion. These psychologists date their scientific origins all the way back to Francis Galton, who argued that human temperament and character were "durable realities, and persistent factors of our conduct."[1]

Situation psychologists, on the other hand, claim the environment drives our personality far more than personal traits. They believe that culture and immediate circumstances determine how we behave, arguing, for example, that violent movies are likely to make people aggressive, regardless of innate tendencies.[2] The situation psychologists trace their origins to an equally impressive founder, Adolphe Quetelet, who famously claimed that "society prepares the crime and the guilty are only the instrument by which it is executed."[3]

The famous obedience study by Yale psychologist Stanley Milgram is a classic situationist experiment.[4] In this study participants were told to deliver electrical shocks (ranging from 15V to a potentially fatal 450V) to a person located in another room every time the individual gave an incorrect answer to a question. Unbeknownst to the participants, the people in the other room were actors and were not actually receiving electrical shocks. Milgram wanted to know: How far could people be pushed to harm another person if given orders by an authority figure? The results were alarming: 65 percent of people administered the full 450V shock even when the person in the other room begged them not to, cited heart problems, or simply stopped responding.[5] According to situationists, the results of this study proved that a strong situation influences the behavior of most people, even compelling them to acts of cruelty.

Throughout the twentieth century, the trait and situation theorists battled it out in the halls and laboratories of academia, but by the 1980s the trait psychologists emerged as the undisputed victors.[6] While the situation psychologists were able to predict, on average, how *most* people would behave in a situation, they could never predict how any particular *individual* was going to behave. They could, for instance, predict that a majority of people would deliver an electrical shock to an innocent stranger if commanded to by an authority figure, but not whether Mary Smith from Cincinnati was more likely to do so than Abigail Jones from Tallahassee.

In contrast, the trait theorists did a better job of predicting the behavior of any given individual—on *average,* at least. They also produced something far more useful to business: personality tests. Today, twenty-five hundred different kinds of trait-based assessments are administered each year to employees.[7] For example, eighty-nine of the Fortune 100 companies, thousands of universities, and hundreds of government agencies use the Myers-Briggs Type Indicator (MBTI) test, which assesses four dimensions of personality and categorizes people into sixteen distinct types.[8] Salesforce.com, meanwhile, relies on the Enneagram personality test to evaluate applicants, a test that assigns people one of nine numbered personality types ("Type 8," for instance, is a "Challenger").[9] These and other tests are all part of a half-billion-dollar industry that is dedicated exclusively to measuring and classifying our personality traits.

But perhaps the greatest reason for the success of trait theory is that it seems to jibe with our private sense of ourselves—and others. When confronted with the Myers-Briggs, for instance, we tend to instinctively map our personality onto its structure, quickly deciding if we are introverts or extroverts, thinkers or feelers, judgmental or perceptive. Similarly, if asked to describe the personality of our best friend—or our worst enemy—we would most likely offer a list of their prominent traits. They are helpful, optimistic, and impulsive, we might conclude, or perhaps aggressive, cynical, and selfish. Similarly, if I asked you to point out a few introverted colleagues, I suspect you would have little difficulty providing names.

Tests that score us on a set of traits are popular because they satisfy our deep-seated conviction that we can get to the heart of a person's "true" identity by knowing those traits that define the essence of that person's personality. We tend to believe that, deep down in the bedrock of a person's soul, someone is essentially wired to be friendly or unfriendly, lazy or industrious, introverted or extroverted, and that these defining characteristics will shine through no matter what the circumstances or task. This belief is known as *essentialist thinking*.[10]

Essentialist thinking is both a consequence and a cause of *typing*: if we know someone's personality traits, we believe we can classify them as a particular type. And if we know someone belongs to a particular type, we believe we can form conclusions about their personality and behavior. That's what happened to me in seventh grade, after I started a spitball fight in English class. The incident (rightly) earned me a trip to the school counselor's office, and since it was not my first visit, I was required to complete an aggression questionnaire that determined I ranked near the 70th percentile. My parents were called to the school where the counselor informed them that, in his opinion, I was an "aggressive child" and patiently laid out the evidence: my spitballs, a fight earlier in the year, and the most damning evidence of all—the questionnaire results.

The counselor believed that aggressiveness was something essential about my character, a defining feature of who I was, and he understandably presumed this knowledge allowed him to make predictions about me. He recommended that I see a psychologist, warning that aggressive children usually struggle in school and are often not suited for the pressures of college. He also informed my parents that I would struggle with authority figures and would therefore have trouble holding down a job if my counseling was not effective. This is, of course, the reason we rely on essentialist thinking to size people up: knowing someone's traits seems to grant us the ability to predict how they will perform in school, on the job, or even (as dating websites insist) as a romantic partner.[11]

But here's the problem: when it comes to predicting the behavior of individuals—as opposed to predicting the average behavior of a *group* of people—traits actually do a poor job. In fact, correlations between personality traits and behaviors that *should* be related —such as aggression and getting into fights, or extroversion and going to parties—are rarely stronger than 0.30.[12] Just how weak is that? According to the mathematics of correlation, it means that your personality traits explain 9 percent of your behavior. Nine percent! There are similarly weak correlations between trait-based personality scores and academic achievement, professional accomplishments, and romantic success.[13]

If our personality and behavior are not explained by a collection of enduring traits, then how *do* we explain our personalities? After all, our behavior is not random—and it doesn't just depend on the situation alone. The reason trait theory, and the essentialist thinking that supports it, does such a poor job explaining human behavior is because it completely ignores the second principle of individuality: the *context principle*.

THE CONTEXT PRINCIPLE

University of Washington professor Yuichi Shoda is one of the top researchers in child development, and one of my favorite scientists in all psychology.[14] Shoda began conducting personality research as a Stanford graduate student in the 1980s, at the height of the academic conflict between the trait theorists and the situation theorists. But though his research thrust him into the middle of the personality debate, he never sided with either camp. Early on he intuited that both approaches were incomplete and, ultimately, misguided.[15]

Shoda approached human personality analytically and systematically, throwing out old assumptions. In doing so, he became convinced that the perennial conflict between trait and situationist theory was holding the field back because both approaches failed to account for what he saw as the true complexity of the individual. Shoda thought there was a third way to think about personality, not in terms of traits or situations, but in terms of the ways in which traits and situations *interacted*. This was no compromise position—if he was right, it would mean that both sides of the venerable personality debate were wrong.[16]

To persuade other scientists of the legitimacy of his theory, he knew he would need a very convincing study, one that gathered a large amount of behavioral data on individuals across a variety of natural settings. It did not seem feasible to study adults with that kind of comprehensiveness, since it would almost certainly require monitoring them all day long, including at their jobs. Instead, he decided to study children at a residential summer camp program in New Hampshire known as Wediko Children's Services.[17]

The kids at Wediko ranged in age from six to thirteen and were mostly from low-income families in the Boston area. Shoda followed 84 children (60 boys and 24 girls) around the camp during every hour of camp activity, for six weeks, documenting their behavior in every location at Wediko except for the bathrooms. To accomplish this enormous undertaking, Shoda relied on a team of seventy-seven adult camp counselors who recorded more than 14,000 hours of observation, an average of 167 hours for each child. The camp counselors also filled out subjective ratings for every child at the end of every hour.[18]

At the end of the summer, Shoda painstakingly sifted through this massive bundle of data by first analyzing each individual child's behavior, and then looking for collective patterns. The results were plain and unmistakable—and a direct blow to essentialist thinking: each child exhibited different personalities in different situations.[19]

Now, in some sense, this is no big surprise, and you might be quick to counter, "Of course we behave differently in different settings!" But think about trait models of personality for a moment. The Myers-Briggs, for example, definitely does *not* say that our traits fundamentally change depending on the setting; as a matter of fact, it says the opposite: that dispositions like whether we are introverted or extroverted influence our behavior no matter the situation. Trait-based personality tests assume that we can be *either* extroverts *or* introverts . . . but not *both*. Yet, Shoda discovered that every child really *was* both.[20]

A girl might be extroverted in the cafeteria, but introverted on the playground. A boy might be extroverted on the playground, but introverted in math class. And it was not the situation alone that was the determining factor: if you picked two girls, one might be introverted in the cafeteria and extroverted in the classroom, while the other might be extroverted in the cafeteria and introverted in the classroom. The way someone behaved always depended on both the individual and the situation. There was no such thing as a person's "essential nature." Sure, you could say someone was more introverted or extroverted *on average*—this was, in fact, exactly what trait psychology amounted to. But if you relied on averages, then you missed out on all the important details of a person's behavior.

Shoda's results directly contradicted the basic tenets of trait theory. Assessing personality *on average* may have been good enough for academics trying to draw broad conclusions about groups of people, but it is not good enough if you are looking to hire the employee best suited for the job or to deliver the most effective counseling to a student, and it is not nearly good enough for making decisions about *you*. Defining yourself as "generous" or "stingy" does not do justice to the fact that you donate money to a struggling non-profit organization with a sense of mission, but not to your well-endowed alma mater. However, Shoda's results also repudiated situation theory, since his data demonstrated that any given situation affected each person differently. Not surprisingly, when personality psychologists learned about Shoda's results, many reacted the same way psychometricians did when they first heard about Peter Molenaar's ergodic switch: by accusing Shoda of promoting anarchy.

Shoda seemed to be suggesting that nothing about people's personalities was consistent, that their behaviors were a constant whirlwind, shifting randomly from place to place. What were personality theorists supposed to model if traits were no longer stable? But Shoda wasn't undermining the concept of personality—rather, by placing the person and context together, he was giving it life. Shoda demonstrated that, in fact, there *is* something consistent about our identity—it just wasn't the kind of consistency anyone expected: we are consistent *within a given context*. According to Shoda's results (as well as a great deal of subsequent research), if you are conscientious and neurotic while driving today, it's a pretty safe bet you will be conscientious and neurotic while driving tomorrow. At the same time, what makes you uniquely you is that you may *not* be conscientious and neurotic when you are playing Beatles cover songs with your band in the context of your local pub.

Shoda's research embodies the second principle of individuality, the *context principle,* which asserts that individual behavior cannot be explained or predicted apart from a particular situation, and the influence of a situation cannot be specified without reference to the individual experiencing it.[21] In other words, behavior is not determined by traits or the situation, but emerges out of the unique interaction between the two. If you want to understand a person, descriptions of their average tendencies or "essential nature" are sure to lead you astray. Instead, you need a new way of thinking that focuses on a person's context-specific behavioral signatures.

IF-THEN SIGNATURES

Shoda summarized his trailblazing findings in his aptly titled book: *The Person in Context: Building a Science of the Individual.*[22] In it, he provides an alternative to essentialist thinking he calls "if-then signatures."[23] If you want to understand a coworker named Jack, for example, it is not particularly useful to say "Jack is extroverted." Instead, Shoda suggests a different characterization: *IF* Jack is in the office, *THEN* he is very extroverted. *IF* Jack is in a large group of strangers, *THEN* he is mildly extroverted. *IF* Jack is stressed, *THEN* he is very introverted.

An example from Shoda's study illustrates the practical value of knowing someone's if-then signatures. When evaluated using standard aggression questionnaires, two boys at Wediko exhibited almost

identical levels of aggression, which, interpreted through the lens
of essentialist thinking, would lead you to assume that their future
outlooks were similar and that they required similar forms of intervention. Yet Shoda's data revealed a hidden distinction—a distinction
that makes all the difference for understanding these children. One
of the boys was aggressive around his peers, but docile around adults.
The other boy was only aggressive around adults, but docile around
his peers. The aggressiveness of each boy was markedly different, and
yet these crucial differences were erased using a trait-based score.
Aggression was not the "essence" of each boy's personality—rather,
there were situations where each boy was aggressive, and situations
where he was not. There was a real cost to ignoring contexts and simply tagging each boy with the same averagarian label.

Consider the following figure, which depicts the if-then signatures for aggressiveness in two boys who are based on the boys in
Shoda's study.

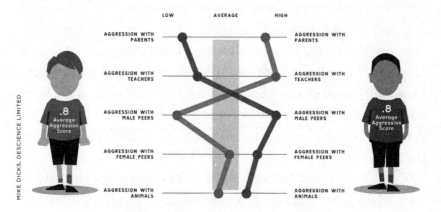

IF-THEN SIGNATURES FOR AGGRESSION

When I first read Shoda's study, I thought back to my experience
when my school designated me an "aggressive child." I recalled that
when my grandmother heard this verdict she refused to believe it,
telling my parents, "He's always so nice at my house!" This was not

grandmotherly obliviousness. I really *was* nice when I was around her. My aggressiveness was triggered by very specific contexts, such as when I was being bullied. In the class where I got in trouble for shooting spitballs there were three bigger kids who liked to push me around. I tried to avoid them outside of class, but within class I often reacted to their presence by becoming the class clown, since I thought if I could make them laugh, they would be more likely to ignore me. It usually worked, though it also earned me a trip to the counselor's office.

If the school administrators (who I genuinely believe cared about me) had attempted to understand the context of my behavior, perhaps they could have helped me, instead of labeling me as aggressive, instead of consigning me to the troubled realm of the "problem child." If they had tried to gain insights into why I was misbehaving *in that context* perhaps they could have intervened—by talking to the teacher or moving me to a new class—instead of presuming they understood something essential about my character.

Later, when I managed to attend Weber State University, I used my knowledge of my if-then signatures to change the way I approached my classes. One invaluable thing I did from the beginning was to avoid classes where I knew other students from my high school. I knew that particular context would lead me to behave like the class clown, and I knew I would never be successful in college as the class clown.

Similarly, I knew that I responded well to certain teaching styles. I especially liked teachers who challenged students to think for themselves and argue over ideas, while I tended to get frustrated and disengaged from teachers who felt that the facts were known and it was our job to sit there and digest them. So at the beginning of each semester I signed up for six courses and attended at least one session of each. If there was a kid I already knew, or if the teacher's style was a bad fit, then I simply dropped the class.

Knowing how I behaved in certain contexts allowed me to make better decisions as a college student and beyond.

ARE YOU HONEST OR DISHONEST?

It's not hard to accept our if-then signatures when it comes to personality—to accept that we might be simultaneously aggressive with some people and nice and quiet with others, or that our introversion or extroversion is specific to whatever situation we find ourselves in. But what about honesty? Loyalty? Kindness? Aren't these inherent to our character? Or is character, too, a changeable, contextual quality?

For a long time, it was believed that people's character is burned into their nature. If we learn that our neighbor's son was caught shoplifting candy at the local convenience store, we instinctively presume he is going to steal other things, too. We certainly would not leave him alone in our home. We might even be inclined to believe that he has some defect of moral fiber that will inevitably drive him to not only commit further acts of thievery, but will most likely lead him to behave in other unscrupulous ways, such as cheating at school and lying to adults.

This view, as it turns out, is wrong. Character is no different than any human behavior: it is meaningless to talk about it in the absence of context. In an era when we are still having heated debates about how to instill moral traits like empathy, respect, and self-control in our children, when we believe that someone is either *honest* or *dishonest* with nothing in between, the idea that each of these important qualities is characterized by a highly individualized if-then signature can seem provocative. And yet, the knowledge that character is contextual is nothing new.

One of the earliest large-scale scientific investigations of character was conducted in the 1920s by a psychologist and ordained minister named Hugh Hartshorne.[24] Those were the heady days when schools across America were getting standardized, and a heated debate arose about whether and how schools should teach character.[25] As president of the Religious Education Association, Hartshorne personally believed that religious education was the best means for inculcating moral values in young people. But, as a scientist, he also knew that before he could advocate any particular approach, he first needed to conduct research to clarify the nature of character.

Hartshorne's team examined 8,150 public school students and 2,715 private school students between the ages of eight and sixteen years. Each student was placed in twenty-nine different experimental contexts that included four different situations (school, home, party, and athletic contest) and three possible acts of deception (lying, cheating, or stealing). Each context was manipulated to have two conditions. In the first condition (the monitored condition), there was no way for students to behave dishonestly. For instance, while taking a test at school they were watched closely by the teacher, who then graded their answers. In the second condition (the unmonitored condition), students were led to believe that any deception they committed would not be detected. For instance, after taking a test at school, they were given the opportunity to grade their own test alone in a room, but Hartshorne inserted a hidden carbon sheet beneath the test to detect whether students changed answers for a better score. For any given context, the difference between a student's behavior in the monitored and unmonitored conditions provided a measure of the student's honesty in that context.[26]

When he began the study, Hartshorne viewed honesty through the prism of essentialist thinking, expecting that each individual student would be either virtuous or unvirtuous. But that was not what he found at all. Instead, students showed little consistency in their virtue. A child who cheated in scoring her own test might be honest when keeping score at a party game. A child who cheated on a test by copying another student's answers might not cheat when he scored his own test. A child who stole money at home might not steal money at school. Honesty, it turned out, was contextual.[27]

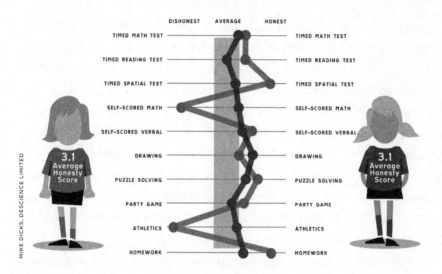

IF–THEN SIGNATURES FOR HONESTY

To get a sense of what Hartshorne found, take a look at the if-then signatures for two eighth-grade students from his study. Each student had a similar average honesty score. The student on the right, with one exception, consistently showed the same level of honesty, regardless of opportunities to cheat. Hartshorne stressed that this student was a true rarity: out of the 10,865 students in his study, she was the most consistent performer by far, with an honesty profile that is much flatter than anyone else's.[28] The student on the left, meanwhile, possesses a decidedly different if-then signature. That student's behavior varied wildly across contexts, all the way from the most meticulous honesty to the most egregious cheating. Yet if you take an essentialist view of character, you would conclude there is no difference between these students—they are equally honest, on average. The context principle, however, shows us that such a view is wrong because it ignores the individuality of each student.

When the public learned of Hartshorne's results, there was widespread shock and outrage. "There has been no more disconcerting theory for parents and teachers generally than the doctrine that moral behavior is specific and conditioned to a large degree by the external situation," Hartshorne declared in response. "If Johnny is honest at home and you should remark that he cheats in his school examinations, his mother is more apt to be incredulous. However repugnant to popular opinion, the doctrine of specificity would seem to be well established . . . honesty, charity, cooperation, inhibition, and persistence are particular habits rather than general traits."[29]

Things have not changed very much, and today parents and teachers still want to believe that moral fiber is a personal trait and not dependent on the situation. Take self-control. Parents are bombarded with studies and books that claim that self-control is the key to our children having a successful life.[30] One of the most famous studies cited to support the importance of self-control—and arguably the most famous psychology study of our generation—is the so-called "marshmallow study."

The general framework for the marshmallow study has been replicated many times.[31] In the most common version of the study, an adult presents a child, usually between three and five years of age, with a marshmallow and a choice. They can eat the marshmallow immediately, or wait fifteen minutes and receive a second marshmallow. The adult leaves the room. The length of time the child lasts without eating the marshmallow is taken as a singular measure of their self-control, ranging from low to high.

The marshmallow study was invented more than forty years ago by a Columbia University psychologist named Walter Mischel.[32] The popular influence of the study really exploded, however, when Mischel and our friend Yuichi Shoda followed up with the participants from the original studies years later and found that, on average, the participants who exhibited the highest self-control as children tended to be the ones who were better socially adjusted, and had achieved greater academic success as adolescents.[33]

This set off nothing less than a self-control craze that stretched across science, parenting, and education. Neuroscientists sought out the "self-control" structures in the brain that enabled kids to resist the temptation of eating marshmallows,[34] child psychologists developed programs that parents could use to increase self-control in their sons and daughters,[35] and educators rushed to promote new forms of character education believed to help boost self-control.[36] Pundits and the media suggested that weak-willed children who could not wait patiently for additional marshmallows were at grave risk for failure in life.[37] Of course, the entire marshmallow-fueled furor was based on the implicit assumption that self-control was an essentialist trait.

"It was very ironic that everyone used the study to support the trait perspective and promote character education," Shoda told me. "Because Walter [Mischel] was fighting against that his whole career. In actuality, we were trying to show that kids can enhance their control over situational pressures through if-then strategies."[38]

The context principle reminds us that self-control does not exist apart from a particular situation, and one person who recognized that context was missing from the popular accounts of the marshmallow test was a scientist named Celeste Kidd.[39] Now an Assistant Professor of Brain and Cognitive Sciences at the University of Rochester, Kidd was working as a volunteer at a homeless shelter when she first heard about the marshmallow studies. "There were many kids staying at the shelter," she told me. "If a child got a toy or candy there was always the real risk that another kid would simply grab it, so the safest and smartest thing to do was to either keep it hidden or eat it as quickly as you could. So when I came across the marshmallow studies my immediate reaction was that every kid who stayed in the shelter would eat the marshmallow right away."[40]

Kidd conducted her own version of the marshmallow study, with a crucial twist: she placed one group of children in a "reliable" situation and another group in an "unreliable" situation. Before the marshmallow test began, the kids in the unreliable situation encountered an adult who did not keep his word—for example, during an art project the adult promised the child that if she waited for a little while he would bring her a new set of art supplies to replace her container of broken and well-worn crayons. After a few minutes, he returned empty-handed. The kids in the reliable group, meanwhile, encountered an adult who delivered the new supplies exactly as promised.[41]

The kids in the reliable situation behaved pretty much like the kids in previous marshmallow studies: a few kids gave into temptation quickly, but about two-thirds of them managed to wait all fifteen minutes—the maximum time. Things were quite different for the kids in the unreliable situation. Half of them devoured the marshmallow within the first minute after the adult departed. Only one child lasted long enough to get a second marshmallow.[42] Self-control feels like some kind of essential trait, but Kidd helped show that it, too, is contextual.

MATCHING TALENT WITH CONTEXT

The popularity of the marshmallow test and its conclusion that self-control is the key to success shows that the one domain where society remains most bound to essentialist thinking is in our attitude toward ability, talent, and potential. We imagine that these are essential qualities—that individuals either possess them or they don't, that circumstances might have some minor influence over something like talent, but the circumstances don't determine or create talent.

Nowhere is that reflected more than in how we hire employees. When it comes to finding the best person for the job, all the systems of our business world are set up to ignore the context, and it starts with the most essentialist hiring tool of them all: the job description. A typical job description for a director of marketing position might include a "key qualifications" or "required skills" section that looks something like this:

- Must have 10 or more years of progressive marketing and sales management experience.
- Bachelor's degree is required; master's degree is preferred.
- Must possess exceptional communication, strategy, and leadership skills.
- Must be a pro at multichannel marketing and managing affiliate programs.

Hundreds of thousands of businesses every week post similarly drafted job descriptions to attract candidates for open positions. Recruiters list the experience, skills, and credentials an employer is looking for, then filter out applicants who don't meet these criteria and select the best candidate among those who are left. At first glance, this seems like common sense: candidates either have certain skills or abilities, or they don't; either you are a "good communicator" or you are not; either you are "a pro" at something like multichannel marketing, or you are not. Of course, the reason it's so difficult to see what's wrong with the approach is because we've been duped into essentialist thinking.

Instead of focusing on the "essence" of the employee, the context principle suggests that a better starting point is to focus on the performance that we need the employee to perform, and the context in which that performance will occur. One person who has pioneered exactly such an approach is Lou Adler, founder of the Lou Adler Group, and one of the most influential recruiting and hiring consultants around.[43]

Before switching to a career in recruiting, Adler designed missiles and guidance systems for an aerospace manufacturer. As a result, he approached the practice of finding and selecting employees with the mindset of an engineer. "One day it just hit me: Once you see how performance depends on context, and how recruiting should be focused on matching individuals to optimal contexts, it just seems like common sense," Adler explained to me. "But it turned out to be really hard to get companies to implement common sense."[44]

Inspired by his context-focused vision for the workplace, Adler developed a new way to recruit and hire employees that he calls "performance-based hiring." Instead of describing the *person* they want, Alder tells employers to first describe the *job* they want done.[45] "Companies always say they want a good communicator. That's one of the most common skills you see on a job description," Adler explained to me. "But there's no such thing as an all-around 'good communicator.' There are many different kinds of communication skills you might need in a particular job, and there's no such thing as someone who's good at all of them." For a customer service rep, good communication is asking questions to understand a customer's problem. For an accountant, it might be explaining to a senior executive how a shortfall in sales affects earnings. For an account executive, it might be leading a full-day presentation to a buying committee. Adler's revelation was that all these contextual details for the performance of "good communication" really mattered.[46]

The Adler Group has helped more than ten thousand hiring managers at businesses ranging from start-ups to Fortune 500 companies switch to performance-based hiring.[47] One client who raves about the impact that performance-based hiring has had on his firm is twenty-five-year-old wunderkind Callum Negus-Fancey, the founder of London-based Let's Go Holdings.[48] The company quickly made a name for itself as "brand advocacy specialists" for media and tech companies and, as a result, Let's Go experienced very rapid growth in its first three years.[49]

"At first, we really didn't know what we were doing when it came to hiring, so we just used the traditional job description approach," Callum told me. "We needed someone to run a marketing team, and we hired someone who matched our generic job description. He had a lot of impressive experience, but his experience was in big corporations, and when he started working for us, a fast-moving start-up, he simply didn't fit in at all. It was a disaster."[50]

That's when Callum heard about performance-based hiring and asked Adler to help him find a new human resources manager. "Adler showed us that what really mattered was selecting someone with success performing in similar contexts to the ones at Let's Go," Callum told me. In this case, Adler's model ended up identifying a very counterintuitive prospect: a pharmacist from Belgium. "Thierry Thielens wasn't British, and he had never done anything in human resources before," Callum recalled. At first, Callum was skeptical, but Adler explained that the pharmacist's previous performance and the conditions he had worked under (such as quickly learning to manage fast-changing staff through a series of new situations), were almost identical to what they needed him to do at Let's Go. So Callum hired him. "Today, he's one of the most important people in the company," Callum told me, "and we would have never considered him if we simply looked at job descriptions."[51]

The human resources industry was born out of Taylorism, with personnel departments tasked with looking for average employees to fill average jobs. Essentialist thinking was fundamental to the mindset from the beginning, and in many ways remains so today. "Companies always lament there's a shortage of talent, that there's a skills gap," Adler told me. "But really there's just a thinking gap. If you spend the effort thinking through the contextual details of the job, you're going to be rewarded."[52] Companies that apply the context principle—companies that attempt to match the if-then signatures of candidates with the performance profiles of the positions they are trying to fill—will end up with more successful, loyal, and motivated employees. For our part, we will have the chance to enjoy a career that matches who we really are.

But a better career match is not the only thing that the context principle opens up for us. It also presents us with a better map for understanding ourselves as well as other people and *their* talents, abilities, and potential. And this deeper understanding of who we are and how we interact with those around us is at the heart of our personal and professional success.

KNOWING OTHERS FOR WHO THEY ARE

The context principle challenges us to think about ourselves and others in a way that is counter to how we've been taught to think about personality most of our lives. It's natural that many people might resist letting go of the idea that, deep down, we must surely possess some kind of enduring, essential traits. Most of us believe that, when it comes right down to it, we are optimists at heart—or cynics. That we are nice—or rude. That we are honest—or dishonest. The idea that who we are changes according to the circumstances we find ourselves in—even if those changes are unique to our own self—seems to violate the fundamental tenet of identity: to us, our personality *feels* stable and steadfast.

We feel this way because our brain is exquisitely sensitive to context and automatically adapts to the situation we find ourselves in. When we are extroverted at a friend's party, our brain instinctively compares our behavior to experiences in similar contexts and concludes that we are acting as expected: we are an extrovert, or at least we are at parties. At work, on the other hand, we might consider ourselves introverted, since our brain remembers that we usually behave in a low-key manner around our coworkers. If our personality feels stable and steadfast, it's because it *is* stable and steadfast—within a given context. Astrologers figured this out long ago, which is why horoscopes often seem persuasive—if the astrologer informs us that Leos are sometimes shy, well, we are *all* shy sometimes. It just depends on the context.

Other people's personalities seem stable to us, however, for a different reason: we tend to interact with most people within a narrow range of contexts. We might know a colleague solely at work, for example—not at home with his family. Or we go out shopping and drinking with a friend on weekends, but never see her in the boardroom. We spend time with our children at home, but rarely see them at school or with their friends. Another reason people's behavior feels trait-like is that *you* are a part of their context. Your boss might think you are a timid person when you know that you are only timid around *her;* at the same time, we might think our boss is overbearing and arrogant, even though she might be only behaving that way around *you.* We simply do not see the diversity of contexts in the lives of our acquaintances or even those closest to us and, as a result, we make judgments about who they are based on limited information.

Breaking free from essentialist thinking, and becoming aware of contextual if-then signatures, can give us an incredible advantage in our personal and professional lives. On a personal level, it helps us more easily recognize the situations where we shine, which allows us to make better decisions. For example, you may excel as part of a collaborative team, but struggle in a context that is more isolated and individualistic, so when offered a big promotion that requires you to work independently from home 90 percent of the time, you might decide to decline because you recognize that, regardless of its benefits, the job does not fit your if-then signature. Conversely, the context principle also helps us identify situational factors that might lead us to behave in negative or self-sabotaging ways, and change or avoid those factors.

In many ways it's not hard to develop awareness of the contexts where we are ourselves successful, and the contexts where we struggle. The hard part is to think about other people's if-then signatures. Essentialist thinking still pervades every aspect of our social lives, and it is hard to resist the pull of false certainty. That's the challenge for all of us—and where the context principle may offer its biggest benefits. Each time we find ourselves thinking someone is neurotic, aggressive, or aloof we should remember that we are only seeing them in one particular context.

Understanding the if-then strategies of others is especially important when we find ourselves entrusted with helping them succeed—as their manager, parent, counselor, teacher, and so on. When we are acting in that capacity, the context principle allows us to deal more productively whenever we see our child, employee, student, or client engaging in negative behaviors we want to change. Instead of asking why they are behaving in that way, we can reframe the question in terms of context and ask ourselves, "Why are they behaving that way *in that context?*" When we see behaviors we think are bad, we can withhold from responding until we first find an example where their behavior *isn't* true (for example, my aggressive behaviors were true in art class, but not true with my grandmother). Or we can follow Celeste Kidd's lead—she told me that any time she finds herself judging someone based on behaviors that strike her as insensitive or irrational, she stops herself, takes a step back, and tries to imagine a set of circumstances that would make the behavior rational and sensible. Most of the time, she realizes that she was projecting her own context onto the other person instead of appreciating his.

Even if we are not entrusted to help others succeed, remembering that we only see others we interact with—like a coworker or boss—in a single context can help us to be more compassionate and understanding with others. If we could see that "difficult" coworker in all her contexts, we might find her to be a devoted friend outside the office, a caring sister, a loving aunt to her nieces. It's then harder to judge that coworker, to reduce her to a singular unflattering personality trait and in the process strip her of what makes her human—her complexity. Remembering that there is more to that person than the context that finds both of us together in that moment opens up the door for us to treat others with a deeper understanding and respect than essentialist thinking ever allows us to. And that understanding and respect are the foundation of the positive relationships that are most likely to lead to our success and happiness.

CHAPTER 6

WE ALL WALK
THE ROAD LESS TRAVELED

One of the most important milestones infants attain in life is standing on their own two feet. For parents, our child's simple act of learning to walk gets bound up in all our hopes and dreams about her future, our yearning to be assured that she is going to be normal, healthy, and *successful*. As we watch our child struggle to pull herself across the floor and lift herself up, we anxiously compare her progress to the prescribed norms. We pay careful attention to whether she sat up at the *right* age and whether she crawls the *right* way. If our daughter lags behind a milestone, we fear it might signal a more serious problem, or worry that she will be burdened by her slow nature throughout her life.

My friend's son recently began crawling in a seemingly unusual fashion: by lying on his side and dragging himself forward with his hands, his immobilized hips and legs sliding along the floor like a little merman. My friend whisked his son off to their doctor, fearing that this aberrant behavior suggested that his son's legs—or, heaven forbid, his brain—had failed to develop correctly. We might chuckle knowingly at such an overreaction, but at the same time every parent understands. Many of us—not just my friend—instinctively regard deviation from the normal pathway as an unmistakable signal that something is *wrong*.

We have already seen how averagarian thinking dupes us into believing in "normal" brains, bodies, and personalities. But it also dupes us into believing in normal pathways—the idea that there is one right way to grow, learn, or attain our goals, whether that goal is as basic as learning to walk or as challenging as becoming a biochemist. This conviction stems from the third mental barrier of averagarianism: *normative thinking*.

The key assumption of normative thinking is that the right pathway is the one followed by the average person, or at least the average member of a particular group we hope to emulate, such as successful graduates or professionals. Just as we believe the countless pediatricians and scientists who have told us that there are ordained milestones for child development, for walking, talking, reading, and everything else.[1]

We owe our sense that there's one right pathway in large part to Frederick Taylor, Edward Thorndike, and their disciples. Taylor laid the foundation for the idea of a standard career track within hierarchical organizations: the average person started as a manager trainee, then got promoted to manager, then department head, then vice president over a division, and so on. His management ideas and his belief that there was "one right way" to accomplish any task in the industrial process helped determine the duration of a workday and workweek—temporal norms originally designed to maximize factory efficiency but which today serve as nearly invisible pacesetters for all aspects of our personal and professional lives.[2]

Taylor's standardization of factory time also inspired the inflexible pathways of our educational system developed and implemented by Thorndike and the educational Taylorists.[3] Our schools still follow the same rigid march through time as they did a century ago, with fixed class durations, fixed school days, and fixed semesters, proceeding through the same unyielding sequence of "core" courses, all of which ensure that every (normal) student graduates from high school at the same age with, presumably, the same set of knowledge.

When you put together a normal educational path with a normal career path, you end up with a normal pathway through life. If you want to become an engineer, you must spend twelve years in school, then four years in college, then take a job as a junior engineer, then hopefully get promoted to senior engineer, project manager, department head, and VP of engineering. In my own profession of academia, the normal pathway is also ordained: school, college, graduate school, postdoc, assistant professor, associate professor, full professor, department chair.

Our shared belief in normal pathways of achievement compels us to compare the progression of our own lives against these average-based benchmarks. The normal time it takes to reach a milestone (such as crawling) or a career goal (like running our own marketing agency) is embedded in our mind like an ever-present stopwatch. If our child starts crawling later than normal, or our former classmate makes director of marketing ahead of schedule, then we often feel like we (and our child) are falling behind.

If we hope to overcome the mental barrier of normative thinking, the first step is to see human pathways of development as they really are.

THE PATHWAYS PRINCIPLE

The act of walking is so universal and so deeply human that it seems almost self-evident that it must develop through a well-defined set of fixed stages—a normal pathway. For almost sixty years, leading researchers and medical institutions agreed, insisting that children crawl, stand, and walk according to a normal developmental timetable. These authorities endorsed a sequence of age-specific milestones that a "typical" child was expected to progress through, based on average ages obtained from large samples of children.[4] The presumption that there *must* be a normal pathway to walking seemed so intuitive and obvious it was almost never challenged. But one person who did was a scientist named Karen Adolph.[5]

Adolph learned the importance of focusing on the individuality of children from her mentor Esther Thelen, the scientist who solved the mystery of the stepping reflex. Adolph has applied this same perspective to her pioneering work on infant development, including crawling. In one study, she and her colleagues tracked the development of twenty-eight infants from before they crawled until the day they walked, examining the data using the "analyze, then aggregate" method. Adolph discovered there is no such thing as a normal pathway to crawling. Instead, she found no less than twenty-five different pathways infants followed, each with its own unique movement patterns, and *all* of them eventually led to walking.[6]

The normal pathway dictated that children should follow certain stages (like rolling onto the belly or moving arms and legs in parallel motions) in a certain order. But Adolph found that some infants exhibited multiple stages simultaneously, or went back and forth between stages, or simply skipped stages altogether.[7] For example, though it was long believed that "belly-crawling" was an essential stage of crawling that infants passed through on their way to walking, almost half of the infants in Adolph's study never belly-crawled at all.[8]

When I first came across Adolph's research, I recalled how my own son walked before he crawled, provoking an irrational surge of pride—Look! My son is going to be an Olympic gymnast!—that abruptly turned to grave concern when, two months later, he "back-tracked" into crawling. But Adolph's research shows that our biology does not compel us to follow a predetermined blueprint. As she explained to me, "Every baby solves the problem of movement in her own unique way."[9]

Even more provocative, it appears that not only are there many ways to learn to crawl, but also crawling itself may not be a universal and necessary step along the path to walking. The idea that crawling is an indispensable stage that precedes walking is a cultural artifact—the result of averaging together the behavior of a highly unusual sample of children: the children of industrialized Western society.

In 2004, anthropologist David Tracer was studying the aboriginal Au tribe in Papua New Guinea when he was struck by an odd revelation: even though he had been observing the Au for twenty years, he had never seen an Au baby crawl.[10] Not one. They did, however, all go through what he called a "scoot phase," where they would shuffle their bottom along the ground in an upright position. Tracer wondered why their pattern of motor development appeared so different from the normal pathway dictated by Western science.[11]

He decided to investigate further, following 113 infants from birth to thirty months of age, documenting their daily interactions with their caregivers as well as assessing them using standardized tests of infant motor development. He discovered that Au caregivers interacted with their babies in a fundamentally different way than Western caregivers. Au babies were carried upright in a sling nearly 75 percent of the time, and in those rare instances where infants were on the ground their caregivers did not permit them to lie face-down. There was a good reason for this restriction: the Au knew their babies would likely pick up fatal diseases and parasites if they made extensive contact with the ground.[12]

In the West, we take it for granted that the floor of our home is relatively free of dangerous germs, and so never question whether crawling is an essential stage in motor development. It is a powerful reminder that far too often we interpret average patterns of behavior as proof that something is innate and universal, when in fact the patterns might stem entirely from social customs that constrain what pathways are even possible in the first place.

Of course, it's not that aberrant pathways don't exist. There are developmental wrong turns and dead ends; on occasion, kids really do have medical issues that prevent them from moving properly and intervention is required. But these medical issues, like walking itself, are individualistic enough that they cannot easily be understood simply by comparing how far an infant deviates from average development.

Normative thinking—the belief there is one normal pathway— has fooled scientists in many fields, not just child development. Take, as an example, colon cancer, one of the most common and lethal forms of cancer in the world.[13] For several decades, it was assumed that a "standard pathway" dictated how colon cancer formed and progressed, a fixed bio-molecular sequence driven by the unfolding of specific genetic mutations.[14] How did scientists derive this standard pathway? By averaging together findings from a wide variety of individual colon cancer patients.

The notion of a standard colon cancer pathway remained the consensus view among scientists until researchers, armed with more data and powerful methods, began focusing on individual patients instead of averages. They found, to their surprise, that the standard pathway only accounted for 7 percent of actual cases of colon cancer. Instead, researchers discovered there were multiple forms of colon cancer, each with its own developmental pathway—pathways that had been concealed by scientists' belief that there *must* be a standard pathway.[15] The recognition of multiple pathways has led to major breakthroughs in research and treatment, including earlier identification of the disease and the development of more effective drugs that target specific patterns of colon cancer.[16]

Normative thinking also permeates mental health. For a long time, clinicians who treated depression assumed that all patients in cognitive therapy (a common form of psychotherapy) followed a standard pathway to recovery, a pathway obtained from the average of many patients' recovery experiences. According to the standard pathway, patients showed a quick reduction in symptoms, followed by slow gains the rest of the way.[17] This standard pathway is widely used to benchmark progress for patients treated with this approach. However, in 2013, a team of researchers who were focused on studying individual recovery outcomes instead of average outcomes discovered that the average recovery pathway only applied to 30 percent of patients. They also found two alternative pathways to recovery: in one, patients made slow linear progress; in the other, patients showed a dramatic onetime drop in symptoms, then made little recovery after that. It turns out there was nothing optimal or even "normal" about the average recovery pathway.[18]

The fact that there is not a single, normal pathway for *any* type of human development—biological, mental, moral, or professional— forms the basis of the third principle of individuality, the *pathways principle*. This principle makes two important affirmations. First, in all aspects of our lives and for any given goal, there are many, equally valid ways to reach the same outcome; and, second, the particular pathway that is optimal for you depends on your own individuality.

The first point is rooted in a powerful concept from the mathematics of complex systems called *equifinality*.[19] According to equifinality, in any multidimensional system that involves changes over time—like a person interacting with the world—there are *always* multiple ways to get from point A to point B. The second point is derived from the science of the individual, which tells us that, because of the jaggedness and context principles, individuals vary naturally in the *pace* of their progress, and the *sequences* they take to reach an outcome.[20] It is in understanding the *why* that we discover how to leverage the pathways principle to work for us as individuals and as a society.

THE PACE OF EXCELLENCE

If you believe only one pathway exists to achieve your goal, then all there is to evaluate your progress with is how much faster or slower you hit each milestone compared to the norm. Consequently, we bestow tremendous meaning on the pace of personal growth, learning, and development, equating *faster* with *better*. Terms like "whiz kid" or "quick study" reflect our cultural faith that faster means smarter. If two students earn the same grade on a test, but one student finished in half the time, we assume the faster student is the more gifted one. And if a student should need extra time to complete an assignment or to finish a test, the presumption is that he is not particularly bright.

The assumption that faster equals smarter was introduced into our educational system by Edward Thorndike. He believed that the pace at which students learned material was correlated with their ability to retain it, which in turn was correlated with academic and professional success. Or, in his words, "it is the quick learners who are the good retainers."[21] He explained this purported correlation by arguing that differences in learning were a result of differences in a brain's ability to form connections.[22]

Thorndike recommended standardizing time for classes, homework, and tests based on how long it took the *average* student to complete a task as a way to efficiently rank students. Since he equated faster-than-average with smarter-than-average, he presumed the smart students would perform well when given an average allotment of time. On the other hand, since he presumed the dull-witted students would not perform much better no matter how much time you gave them, there was no point in offering more than an average allotment of time, especially since it would only hold back the bright students.[23] Even today, we remain reluctant to grant students extra time to complete tests or assignments, believing that it is somehow unfair—that if they are not fast enough to finish these tasks in the allotted time, they should be appropriately penalized in the educational rankings.[24]

But what if Thorndike was wrong? If speed and learning ability are *not* related, it would mean that we have created an educational system that is profoundly unfair, one that favors those students who happen to be fast, while penalizing students who are just as smart, yet learn at a slower pace. If we knew that speed and learning ability were not related, we would, I hope, go to great lengths to provide students with as much time as they needed to learn new material and complete their assignments and tests. We would evaluate students based on the quality of their outcomes, not the quickness of their pace. We would not rank students based on how they performed on a high-stakes test that must be finished in a fixed amount of time.

The fundamental nature of educational opportunity in our society hinges on the question of how speed and ability are related—and it turns out that we have known the answer for the past thirty years thanks to the pioneering research of one of the most famous educational scholars of the twentieth century, Benjamin Bloom.[25]

In the late 1970s and early 1980s, scholars and politicians in the United States debated whether schools could narrow achievement differences, or whether these were mostly due to factors outside the schools' control, such as poverty. Bloom, then a professor at the University of Chicago, was convinced that schools mattered. He believed the reason many students struggled in school had nothing to do with differences in the capacity to learn, and everything to do with artificial constraints imposed on the education process, especially fixed-pace group instruction—when a curriculum planner determines the pace at which the whole class should be learning the material.[26] Bloom argued that if you removed this constraint, student performance would improve. To test this hypothesis, he designed a series of experiments to determine what would happen if students were allowed to learn at their own pace.

Bloom and his colleagues randomly assigned students to two groups.[27] All students were taught a subject they had not learned before, such as probability theory. The first group—the "fixed-pace group"—was taught the material in the traditional manner: in a classroom during fixed periods of instruction. The second group— the "self-paced group"—was taught the same material and given the same *total* amount of instruction time, but they were provided with a tutor who allowed them to move through the material at their own pace, sometimes going fast, sometimes slow, taking as much or as little time as they needed to learn each new concept.[28]

When Bloom compared the performance of students in each group, the results were astounding. Students in the traditional classroom performed exactly like you would expect if you believed in the notion that faster equals smarter: by the end of the course, roughly 20 percent achieved mastery of the material (which Bloom defined as scoring 85 percent or higher on a final exam), a similarly small percentage did very poorly, while the majority of students scored somewhere in the middle. In contrast, more than 90 percent of the self-paced students achieved mastery.[29]

Bloom showed that when students were allowed a little flexibility in the pace of their learning, the vast majority of students ended up performing extremely well. Bloom's data also revealed that students' individual pace varied depending on exactly what they were learning. One student might breeze through material on fractions, for instance, but grind through material on decimals; another student might fly through decimals, but take extra time for fractions. There was no such thing as a "fast" learner or a "slow" learner. These two insights—that speed does not equal ability, and that there are no universally fast or slow learners—had actually been recognized several decades before Bloom's pioneering study, and have been replicated many times since, using different students and different content, but always producing similar results.[30] Equating learning speed with learning ability is irrefutably wrong.

Of course, the conclusion that logically follows from this is both obvious and terrible: by demanding that our students learn at one fixed pace, we are artificially impairing the ability of many to learn and succeed. What one person can learn, most people can learn if they are allowed to adjust their pacing. Yet the architecture of our education system is simply not designed to accommodate such individuality, and it therefore fails to nurture the potential and talent of *all* its students.

Of course, it is one thing to recognize a problem, and another thing entirely to fix it. In the 1980s, when Bloom conducted his research, he acknowledged that it would be prohibitively complex and expensive to convert our fixed-pace standardized education system into a flexible-paced one.[31] But the '80s have passed. We live in an era where new, affordable technology can make self-paced learning an accessible reality.

Khan Academy is a nonprofit educational organization that provides, in the words of its website, "a free, world-class education for anyone, anywhere."[32] Today, Khan Academy boasts over ten million users around the world and consists of an extensive set of online modules covering just about any academic subject imaginable, from ancient history to macroeconomics.[33] Perhaps the most notable thing about Khan's modules (besides their zero cost), is the fact that they are entirely self-paced: the software adapts to each student's pace of learning and only proceeds to a new set of material when the student has mastered the present set.[34]

Since Khan records data on each student's progress, it is possible to track the individual learning pathway of each student who uses the modules. The data confirm precisely what Bloom first discovered more than thirty years ago: every student follows a unique pathway that unfolds at his or her own highly individualized pace. The data also confirm that the pace at which any given student learns is not uniform: we all learn some things quickly and other things slowly, even within a single subject.[35]

In his widely viewed 2011 TED talk, Khan spoke eloquently about the relationship between pace and learning: "In a traditional model, if you did a snapshot assessment [of student performance after a fixed period of time], you say, oh, these are the gifted kids, these are the slow kids. Maybe they should be tracked differently. Maybe we should put them in different classes. But when you let every student work at their own pace . . . the same kids that you thought were slow six weeks ago, you now would think are gifted. And we're seeing this over and over again. It makes you really wonder how much all of the labels a lot of us have benefited from were really just due to a coincidence of time."[36]

Why should we care if it takes a child two weeks or four weeks to learn to solve quadratic equations, as long as she can solve them? Why should we care if it takes a dental student one year or two years to learn to perform a root canal, as long as he or she can perform it flawlessly? There are already many domains in life where we do not particularly care how long it takes someone to achieve mastery—we only care that they have mastered it. Driving, for instance. A driver's license does not record how many times you failed the written driving exam, or the age when you finally obtained it. As long as you passed the driving exam, you are allowed to drive. The bar exam is another familiar example: obtaining your license to practice law does not depend on how long it took you to pass the exam, only that you passed.

If every student learns at a different pace, and if individual students learn at different paces at different times and for different material, then the idea that we should expect every student to learn at a fixed pace is irredeemably flawed. Think about it: Were you really not good at math or science? Or was the classroom just not aligned to your learning pace?

WEBS OF DEVELOPMENT

It's not too hard to believe that everyone develops at a different pace, or even that we each progress at different paces in different domains. What can be much harder to accept is the second assertion of the pathways principle: there are no universally fixed sequences in human development—no set of stages everyone must pass through to grow, learn, or achieve goals. The idea of normative stages gained widespread public support in the early twentieth century because of the work of a pioneer in infant studies, an American psychologist and pediatrician named Arnold Gesell.[37]

Gesell believed that evolution designed the human brain to unfold in a certain sequence determined by biological maturation, so that the mind had to learn and adjust to specific things about the world before it could proceed to more advanced stages, with each new stage serving as an essential foundation for the next.[38] Gesell was the first scientist to track the development of large numbers of infants and the first to use the average of their development to describe fixed milestones that he believed represented the normal progression of a typical child.[39]

Gesell found average-based stages everywhere he looked: for example, he identified 22 stages in the development of crawling, including "lifting the head and chest off the ground, pivoting in circles, pulling forward with the abdomen dragging along the ground, hopping forward with the belly alternately on and off the ground, rhythmical rocking on hands and knees, crawling on hands and knees, creeping on hands and feet."[40] He claimed to have identified 58 stages of behavior when playing with a pellet (twenty-eight-week-old toddlers place an open hand over a pellet, according to Gesell, while forty-four-week-old toddlers hold it tightly) and 53 stages of rattle-grasping behavior.[41] He even coined the term "Terrible Twos" and the phrase "he's just going through a stage."[42]

Gesell set up a laboratory at Yale University where he tested babies and gave them "Gesell scores" indicating how their physical and mental development compared to the norm.[43] If a child failed to progress through the proper sequence of stages, parents were often told (or left to assume) that something might be wrong with their child.[44] These "Gesell scores" were also used as a basis for adoption: Gesell believed that he could improve the success of adoptions by matching smart babies with smart parents, and average babies with average parents.[45] Many medical organizations, including the American Pediatrics Society, endorsed Gesell's framework at the time,[46] and today his ideas still form the basis for the "normal" ages for developmental milestones used in many pediatric guides and popular parenting books.[47]

Gesell and nearly a century of stage theorists have viewed development as a kind of immutable ladder, believing that from the moment of birth, we are each predestined to climb this same ladder rung by rung.[48] But starting in the early 1980s, some researchers began to notice that many of the children in their studies did not conform to the prescribed sequences that were thought to be universal. Eventually, the discrepancies between individual development and these supposed normal pathways became so obvious that it created what came to be known as "the crisis of variability" in developmental science.[49]

To resolve this crisis, a new generation of scientists committed to understanding human individuality began to outline an alternative to the notion of developmental ladders. One of these researchers is psychologist Kurt Fischer, a pioneering figure in the science of the individual and the scientist who formally introduced me to its principles.[50] He is also, I might add, my mentor. Throughout his career, Fischer has approached his research from an individual-first perspective[51] and has brought the pathways principle to bear on a wide range of developmental issues, including one of my favorite examples: understanding how young children learn to read.

For decades, scientists and educators assumed that children learned to read single words according to a standard sequence of skills, such as learning the meaning of a word before learning to identify the letters of a word, which in turn occurs before learning to produce words that rhyme with a given word.[52] This "standard" reading sequence was derived from group averages, and Fischer had a hunch this averagarian approach had led scientists and educators to overlook something important about the process of learning to read.[53]

To test this hunch, Fischer and a colleague analyzed the sequence of reading development in first-, second-, and third-graders. By focusing on the sequences of each individual student, instead of focusing on group averages, Fischer discovered there were actually *three* different sequences that children could progress through on their way to learning to read single words.[54] One of these was indeed the "standard" path, and 60 percent of the children followed it. But another sequence, which included the same skills as the first but in a different order, was followed by 30 percent of the children who also learned to read just fine. A third sequence was followed by 10 percent of the children. However, unlike the other two, the children who followed this sequence ended up with significant reading problems. These children had been labeled as slow or disabled, but by recognizing that they were proceeding down a pathway known to be faulty, they could receive targeted forms of intervention and compensatory teaching, instead of being viewed as unintelligent or impaired.[55]

After his research undermined the notion of fixed sequences and helped resolve the crisis of variability, Fischer offered a new metaphor for development that he felt would allow people to break free of the old averagarian one. "There are no ladders," Fischer once told me. "Instead, each one of us has our own *web* of development, where each new step we take opens up a whole range of new possibilities that unfold according to our own individuality."[56]

The pathways principle assures us that just as there are no fixed ladders of development in reading, there are no fixed ladders of development for any other aspect of our lives, including our careers. Consider what it takes to become an accomplished scientist. In academic circles, there is usually an implicit assumption about the standard sequence for success: get through graduate school, get a permanent position at a university or research institute immediately after your Ph.D., and then get a series of rapid promotions and increasingly larger research grants. But in 2011, the European Research Council (ERC), worried about the potential negative effects of "normative bias" on the development of young female scientists, decided to find out whether there really was such a thing as a standard pathway for a scientific career of excellence.[57]

To answer this question the ERC funded a study, headed by Dr. Claartje Vinkenburg at Vrije Universiteit Amsterdam, to examine the career pathways of successful and unsuccessful applicants for two prestigious research grants. Rather than finding a standard pathway for successful scientists, Vinkenburg discovered seven distinct sequences, each of which led to career success.[58]

Vinkenburg whimsically named each sequence after a dance. The "Quickstep" and "Foxtrot" corresponded to the conventional notion of a successful career (rapid promotions in a university or research institute), and about 55 percent of scientists achieved success that way. On the other hand, the "Viennese Waltz" and "Jive" pathways consisted of slow but steady progression, with the "Waltz" reaching an academic ceiling without enough time left on the career clock to complete another move. The "Slow Waltz" featured an extended series of postdoc positions. The "Tango" was the most complicated pathway of them all, featuring a series of movements in and out of science, including periods of unemployment. Scientists who performed the Tango were traditionally viewed as ordinary or weak scientists, yet the ERC research showed that they, too, attained scientific excellence.[59]

"It's important to realize that excellence is in every pattern. There's not one way," Vinkenburg told *Science Careers* magazine. "You can have an excellent research idea while taking care of seven kids or looking after a sick parent—or being in the lab 24 hours a day. It shouldn't matter how you got there."[60]

So often we imagine that a pathway to a particular goal—whether it's learning to read, becoming a top athlete, or running a company—is somewhere *out there*, like a trail through a forest cleared by the hikers that came before us. We presume the best way to be successful in life is to follow that well-blazed trail. But what the pathways principle tells us is that we are always creating our own pathway for the first time, inventing it as we go along, since every decision we make—or every event we experience—changes the possibilities available to us. This is true whether we are learning how to crawl or learning how to design a marketing campaign.

It can be frightening to contemplate this fact, because it suggests that familiar guideposts may actually hinder more than help—and if we cannot depend on familiar guideposts, what can we rely on to know how we're doing? That's why the pathways principle works best if we've already spent the effort understanding our jagged profiles and our if-then signatures, because the only way to judge if we are on the right path is by judging how the path *fits* our individuality.

TAKING THE UNCHARTED ROAD TO SUCCESS

When, several years after I flunked out of high school, I finally began college at Weber State, I got plenty of advice about the normal pathway for college success. Before my first day of classes, I sat down with my academic adviser—assigned to me because he handled those students whose last names started with the letters between *Q* and *Z*—so he could review what courses I should take each semester. I took out my pad and pencil and eagerly began writing down everything he said, thinking to myself, *He knows the system here and his job is to figure out what's best for me.* He looked over my high school record, thumbed his beard, and declared, "Given your history of poor academic performance, it makes the most sense if you take all your courses in the usual order. Since you need to pass the remedial math class, take it right now to get it out of the way, and make sure to take the freshman English class during your first semester."

I was grateful for what I presumed was highly personalized advice. A few hours later, I bumped into another freshman who shared the same adviser. Her background was quite different from mine—she was a well-mannered student from a major high school in Salt Lake City who graduated with an A average. We compared notes . . . and I discovered that my adviser had given her the same recommendations he had given to me—minus the remedial math, of course.

After I got over a surge of annoyance, I thought about my situation more carefully. The normal pathway had not worked out for me in high school, so why in the world should I expect it to work for me in college? I did not blame my adviser—it couldn't be easy to dole out customized advice to hundreds of confused freshman over a period of a few days—but I did make a conscious decision to never blindly accept what he or anyone else told me was the proper educational path to follow. Instead, I would forge my own path based on whatever I knew about my strengths and weaknesses.

First up: remedial math class. Should I take it? No way. Remedial math has one of the highest fail rates in colleges across the country.[61] I knew if I sat in a long, boring math class, I would almost certainly fail, too. I researched alternatives and found out I could skip remedial math if I passed a onetime math test called the CLEP.[62] I knew I could study really hard for the test at my own pace and in my own way, and so for a whole year in my spare time I practiced the specific concepts that would be on the test and ended up doing so well on the CLEP test that I was able to skip every math class up to statistics—which turned out to be one of my favorite classes in college. I even became a teaching assistant for a statistics professor.

I also postponed freshman English until my senior year, because I knew I would find it boring and would probably not do too well if I took it right away. (I was right: it ended up being one of my dullest classes at Weber, but by the time I finally took it, I had built up my study skills and was able to grind through it.) I didn't stop there—I rearranged the sequence of my entire four-year schedule so that I took the most interesting courses in my first two years. One of those was an advanced course on plagues that required several prerequisites I didn't have, but I took it anyway because it seemed like it would hold my attention. It did.

As a freshman, I didn't even consider the school's honor program, and not just because I was a high school dropout. I was sure that honors classes meant extra work, and since I still had to hold down a job to support my wife and two sons, I did everything I could to avoid extra work. When I was a sophomore, however, a friend of mine who was an honors student casually mentioned to me that all they did in class was sit around arguing about ideas. He said this as a half-hearted complaint, but I immediately perked up: sitting around arguing instead of listening to long, dry lectures? How do I sign up? I persuaded the honors program director to admit me (not an easy task, given my dismal high school record and mediocre standardized test scores), and I quickly discovered that my friend was right: honors classes were all essays and discussions instead of the rote memory of facts. It was a perfect fit for me.

I am often asked how I turned things around so dramatically in college. I left high school with a D- average, but I graduated from Weber with straight A's. If you asked me right after I graduated, I would probably have said hard work, trial and error, and a little bit of luck. That's still true, but in the years since then I have thought about it more carefully, especially to figure out how I might use the elements of my own success to help other students who feel like they don't fit in. When I consider the decisions I made that contributed to my college success, every one of them was rooted in the belief that a path to excellence was available to me, but I was the only one who would be able to figure out what that path looked like. And to do that, I knew that I needed to know *who* I was first.

My decisions also demonstrate how the jaggedness principle, context principle, and pathways principle ultimately all work hand in hand. To choose the right path for me—selecting the sequence of classes to take, for example—I had to know my own jaggedness (such as my low tolerance for boredom, as well as my ability to focus with laser intensity on those things that did manage to captivate me), and I had to know about the contexts where I would be performing (avoiding classes with kids I knew from high school, and seeking classes that focused on arguments and ideas). By knowing my jagged profile and my if-then signatures, I was able to choose a unique pathway that suited me best.

When you hear my story, you might think I am a special case. But that is really the whole point of the principles of individuality: we are *all* special cases. Once you understand these principles, you can exert better control over your life because you will see yourself as you really are, not as the average says you should be. I am not saying there are a million paths that will get you to where you want to go—designing a killer app, becoming a showrunner for a hit drama, starting your own company. But I am saying that there will always be more than one pathway available to you and that odds are the best one for you will be the one less traveled. So brave new paths and try unexplored directions—they are more likely to lead to success than following the average pathway.

PART III

THE AGE OF INDIVIDUALS

All organizations are based upon fundamental assumptions about individuals, whether they know it or not.

—Paul Green, Morning Star Company

CHAPTER 7

WHEN BUSINESSES COMMIT
TO INDIVIDUALITY

One of my first jobs after I dropped out of high school was working in the layaway department of a big-box store. If I had to use a single word to describe the attitude of my coworkers, it would be *apathy*. Even my boss, a likable and easygoing woman in her late forties, was as disengaged as the rest of us. When I started I was eager to make an impression, so I came up with what I believed to be a better way to organize the layaway tags so that customer's items would be easier to find. I eagerly shared my idea with my boss and asked if we could try it out.

"Why bother?" she replied with a half-hearted shrug. "Even if it is better, corporate will never allow it." A few more weeks on the job and I understood what she meant. My idea for changing layaway tags might have been a good one, but it would have been a waste of time to pursue it because I was merely a tiny, interchangeable cog in a gigantic, well-oiled machine. There was no incentive for deviating from the prescribed course, even if it might benefit the company. I was expected to do a certain set of tasks, defined by somebody else—nothing more, nothing less.

Each of us was viewed as replaceable, and we were replaced, often—during my six-month stint at the big-box store, about a third of my coworkers left, including my boss. The rampant turnover made it difficult to develop trusting relationships with my fellow employees because I knew that everyone was temporary. The company, however, was built for this constant churn. Management had carefully designed the system to be "employee-proof" so that no individual worker could disrupt the operation of the store. This was the trade-off the company had decided to accept: they got cheap, interchangeable labor that kept their streamlined system humming; the employees, like me, lost of any sense of purpose or engagement.

Employee indifference is not unique to any one company or business sector—it is endemic to most organizations that rely on Taylorist principles of standardization and hierarchical management. Taylorist "planners" or managers make all the important decisions about how to standardize the operations, and the workers implement those decisions, right or wrong. This is one reason why a 2013 Gallup study found that 70 percent of employees reported feeling disengaged from their jobs.[1]

A century of averagarian business models stemming from Taylorism has convinced us that for the system to win, the individual must be viewed as a cell on a spreadsheet—like a disposable Average Employee. This conviction is spectacularly wrong. Throughout the book, I've shared the stories of companies like Deloitte, Google, the Adler Group, and IGN—who have adopted, even if implicitly, the principles of individuality with great results. By abandoning the mental barriers of one-dimensional thinking, essentialist thinking, and normative thinking, these companies have been able to create highly engaged and competitive workforces. It's easy to assume that these companies are in a position to do away with the legacy of Taylor's scientific management because they have vast resources, or because they operate in industries that are uncommonly open to unorthodox ways of doing business (like the tech industry). But applying the principles of individuality is an option available to every business, in every industry, and in every country.

Three companies—a retailer, an IT company in India, and an industrial food manufacturer—demonstrate that even in industries or countries where it seems like following an averagarian model is the *only* profitable way to do things—or, at the very least, the *best* way—applying the principles of individuality can produce just as good if not better results.

THE SECRET TO LOYALTY AT COSTCO

According to its employees, Costco is a great employer. For four years in a row, it has earned a place on Glassdoor's list of Top Places to Work, hitting the number two spot in its 2014 list of Top Companies for Compensation and Benefits—only Google was rated higher.[2] There are good reasons that employees praise the big-box retailer. In 2014, the typical Costco employee earned just over $20 per hour, compared to a retail industry average of $12.20 per hour, and 88 percent of them participate in company-sponsored health care. During the great recession of 2008, as other retailers were laying people off, Costco actually gave its employees a $1.50 per hour raise.[3]

These employee-friendly stats did not happen by accident. They are a direct consequence of the company's philosophy toward the individual. "Investing in individuals is the core of what we do," Costco founder Jim Sinegal explained to me. "It's not just a slogan.

People often say they care about individuals, but it's something they print up for PR, not something they believe in. But our assumption all along has been that if you hire great people, give them good wages, treat them with dignity, and give them an honest path for a career, great things will happen."[4]

One of the ways Costco invests in its employees is by giving them power over their career pathway. Management helps employees develop whatever skills they think may be useful for the company and encourages them to explore job openings at Costco, even for positions in departments much different from their present one. Costco backs up this commitment to employee self-determination by promoting heavily from within the organization. More than 70 percent of Costco's managers started out pushing carts or working behind a register.[5]

One example of an employee who forged her unique pathway at Costco is Annette Alvarez-Peters. She attended a few semesters of community college before starting her Costco journey at age twenty-one as a sales audit clerk in the accounting division of a San Diego store.[6] Then she moved into merchandising, where she held several positions, including receptionist, administrative assistant, reorder clerk, and then assistant buyer, where she was responsible for Blank Media (floppy diskettes and blank tapes), and Telecommunications (telephones and cellular). Since she exhibited a knack for buying, she was promoted to Electronics Buyer, and then Beverage Alcohol Buyer for the Los Angeles division. Finally, in 2005, she ascended to her present position, where she is in charge of buying all wine, beer, and liquor for Costco, a position so influential that she was ranked number four on the Decanter Power List of the most influential people in the international wine industry.[7] Her career pathway at Costco took her from an accounting clerk to a position where her decisions can affect the price of wine at your local restaurant and what grapes are planted in Italy.[8]

Annette's professional path would be unimaginable at many companies, where normative thinking compels managers and human resources departments to lock employees into a narrow career path, or where certain positions are cordoned off for employees who fulfill certain requirements, such as getting an MBA or working a specific number of years in the industry. "Annette doesn't seem on paper like the type of person who should be important for the wine industry, but she is," Sinegal tells me. "People outside Costco often seem puzzled by her career path, but nobody inside Costco was."[9]

Matthew Horst's path at Costco was no less unique than Annette's. Matthew's brother, Chris, wrote an open letter to Sinegal and the president of Costco, Craig Jelinek, in which he explained how Matthew had always been limited in his job opportunities because he had been diagnosed as a person with special needs—limited, that is, until he applied to Costco. Matthew got a job pushing carts in the Lancaster, Pennsylvania, Costco. Since then, Matthew has been promoted several times, carving out a career that he loves in the process. "For his entire life, Matthew has been classified and known by his 'special needs,'" wrote his brother. "Since the day he began at Costco, however, his coworkers and customers have valued him because of his unique strengths."[10] Costco didn't evaluate Matthew by comparing his traits to those of an average employee; they evaluated him by what he brought to his job as an individual.

"Fit is everything," Sinegal explained to me. "We look beyond simplistic ideas like a [college] transcript or things like that for hiring. . . . There are attributes that matter at Costco, like being industrious. But how do you see that on a résumé?" Sinegal recognized early on that the best way to identify talented young people was to recruit students from local colleges to work part-time rather than hiring graduates from famous universities. Costco cultivates long-term talent by identifying those part-time workers who demonstrate that they are a good fit with the Costco environment—and at the same time allowing those students to see what Costco has to offer them.[11]

Of course, Costco's commitment to individuality would not matter if the company could not successfully compete in the retail space, an industry with razor-thin margins and formidable labor costs.[12] But Costco has not only been profitable every year since it went public, it has consistently delivered more profitable returns for investors than Walmart.[13] Over the past decade, Costco has grown at an annualized rate of 9 percent per year, making it the third-largest retailer in the United States today.[14] This financial success is even more impressive when you consider that Costco employees are paid approximately 75 percent more than Walmart employees,[15] plus are given those industry-topping benefits. If Costco spends more on employees than a company like Walmart—known for its efficiency and cost cutting not just through its supply chain but its labor expenses as well—how is it able to remain so competitive?

One reason is employee loyalty. Not only are individual Costco employees much more *productive* than those at competitors like Walmart,[16] Costco employees rarely leave the company. At Walmart the turnover rate is approximately 40 percent; at Costco, the turnover rate is 17 percent and drops to an astonishing 6 percent after an employee has been there for a year.[17] One study found that when you factor the hidden cost of employee turnover that comes from having to hire and train the next round of replacements (conservatively set at 60 percent of an employee's salary), Costco actually pays *less* per employee than Walmart.[18] Paradoxically, Costco is beating Walmart at its own game—the game of efficiency.

"Walmart and Target and many other retailers made a different bet than we did," Sinegal told me. "But once you accept that way of thinking, it's very difficult to come back from it. Walmart has over two million employees, and a turnover rate hovering close to 50 percent annually. That means you are replacing a million people every year. Just think about that."[19]

It is tempting to look at a company like Walmart, whose mastery of Taylorist efficiency has produced one of the largest and most robust enterprises in history, and blame capitalism for how that company views its employees. But there is nothing inherent to capitalism that says employers must build their practices, especially their human resource practices, around averages. Costco plays in a similar arena as Walmart's Sam's Club and yet its executives have figured out how to treat their employees like individuals and *still* make a solid profit.

The difference between the two companies lies in what each company truly values. Walmart adopted a Taylorist mindset treating its employees as statistics, as a column of Average Joes who can easily be replaced. Costco makes a meaningful attempt to understand the jaggedness of its employees, recognizing the importance of matching employees to the specific contexts in which they thrive, and empowering employees to pursue their unique pathways. Costco is a place where a part-time worker can become a vice-president, and an accounting assistant can become one of the most powerful wine buyers on the planet. Its employees, in turn, reward Costco with their loyalty and engagement, which fuels Costco's superior performance, customer service, and bottom-line results.

"You cannot run a company like Costco without thinking about individuals. Period," Sinegal told me. "You can make money the other way, but you cannot create a place where everyone wins."[20]

HOW ZOHO OUTCOMPETES GIANTS

Zoho Corporation is the largest information technology products company in India, and one of the first to compete with industry leaders like Microsoft and Salesforce.com.[21] It has accomplished this feat primarily through its unique attitude toward its employees—

not by paying them as little as possible, but by believing that talent can be found in anyone if you look for it in the right way.

After obtaining a Ph.D. in electrical engineering from Princeton, Sridhar Vembu returned to his hometown city of Chennai in 1996 to start the software company that would eventually become Zoho Corporation.[22] Today, Zoho is an international leader in cloud-based business, network, and IT infrastructure management software, where its products compete head-to-head with Microsoft Office and with Salesforce.com's customer relationship management (CRM) offerings.[23] The company employs over twenty-five hundred people across multiple countries, and in 2014 generated estimated revenue of $200 million.[24]

Zoho is a large and successful company today, but when Vembu was starting out, he simply could not compete with much larger Indian software companies for the "best talent"—those candidates who ranked high on conventional one-dimensional academic metrics. Vembu knew that if he wanted to succeed, he would need to find talent that other people had overlooked. "Much of the Indian tech industry looks for paper qualifications, applying strict GPA thresholds and such before they will even consider a candidate," Vembu told me. "We decided to look at people who don't necessarily meet those thresholds."[25]

One of these people was Vembu's own brother. He lacked any background in computer science, did poorly in school, and many in his own family expected that he would "not amount to much." But Vembu gave him a chance. "He learned how to program, and he turned into a terrific programmer. Watching my brother unexpectedly blossom was a real pivotal moment for me," Vembu explained. "I have always been open to the idea that talent can be found anywhere, but seeing it happen in front of my eyes really made me confident that we would find a lot of overlooked talent."[26]

Vembu's intuition was soon backed up by hard evidence. As Zoho hired more and more people from lesser known schools—or with no schooling at all—Vembu discovered that there was little or no correlation between academic performance (as measured by grades and the perceived quality of the diploma) and on-the-job performance. "The narrow pathway of college didn't seem to be necessary to become successful at something like programming. So then I began to wonder, why does everyone make it a precondition for being hired?"[27]

Vembu's philosophy was very similar to Costco's: both companies believed in hiring candidates from places other than famous schools and giving them a chance to show what they could do. But Vembu took this philosophy a step further. If you believe that talent can be found anywhere, then one way to act upon this belief would be to cultivate the talent yourself. Vembu told me, "While most people are willing to accept the idea that there is a lot of untapped talent out there, they have trouble acting on it."[28]

Vembu acted on it in 2005 by creating Zoho University, a remarkable institution designed to identify and develop students to become successful employees at Zoho—and to equip them with the skills to be successful human beings.[29] What makes Zoho University so extraordinary is that the students who enroll are often from the most impoverished parts of India. Zoho *pays* economically disadvantaged youngsters with little schooling to attend the university, where they learn programming skills, along with math, English, and current events. Vembu created a school that went out and found raw, unproven kids and gave them a shot.

This was a risky bet for Vembu. Though Zoho was growing quickly, it still wasn't established yet and certainly did not enjoy the kind of deep pockets to ensure that it could survive if such a big bet turned into a disaster. But the bet was even riskier than counting on finding talent in the unlikeliest of places: Vembu was so opposed to the values of averagarianism that he decided not to operate the university according to the traditional standardize-and-rank mission of most schools.

Almost all the instruction is self-paced and project based. There are no grades; instead, students get feedback on their projects. "We realize that students learn at their own pace, and you have to respect that," Vembu emphasized to me. "If what you care about is how well students will do in your company over the next decade, you soon realize that fast and slow are useless distinctions to make. There just isn't a relationship between learning fast and succeeding."[30]

After twelve to eighteen months of paid training, every student is offered a job. But the students are not forced to sign contracts and are under no obligation to work for the company when they graduate. As Vembu explained to me, "We really want to give them skills that will allow them to succeed at other jobs, or to start their own company. But most of them end up working for us."[31]

So how has this experiment gone? In 2005, Zoho University had six students and one teacher; in 2014, it had a hundred students and seven teachers.[32] What is most amazing, though, is not the number of students, but the talent that Zoho has discovered through this process: to date, more than 15 percent of the hundreds of engineers at Zoho have come through Zoho University,[33] and some of the earliest students who joined are now high-level managers in the company.[34] The program has been so successful that, in 2015, Zoho decided that within the next ten years it plans to have a *majority* of employees come from the university.

Vembu's commitment to individuality is evident not only in the way that he finds talent through Zoho University, but in the freedom that individual employees are given to develop and grow within the company. For example, Zoho does not define jobs rigidly, and does not assume there is an optimal pathway for individuals to move through the company. "About half of the people we hire want to explore and develop something new. We encourage it," Vembu told me. "We don't have rigid job descriptions because they promote rigid thinking and suddenly you think there is a fixed job for you. If you give people flexible pathways, people evolve into lots of roles they would have never thought they were interested in."[35]

Since Vembu does not agree with evaluating people based on averages, there are no performance reviews at Zoho, no scorecards, and no employee rankings. "Placing a grade or a number on a human being is nonsense. Our philosophy is that if there is a manager that has a concern with a team member, they should have a one-on-one discussion then and there and help them."[36]

Zoho also consciously avoids the dangers of one-dimensional thinking when it comes to building teams. "If you take a product team, say for our word processor, the common idea is to get a team with rock star programmers who had high grades from the best schools. That is a mistake! You should have lots of differences in skills sets and talents; if they are all in the same mold, nobody can shine and it becomes too narrow and monocultural. We've found that mixing up different talents and ages and experience actually produces better products. It runs counter to tradition, but our products speak for themselves."[37]

Vembu is not exhibiting excessive bravado when he speaks of the high quality of his company's products. Salesforce.com was sufficiently concerned with Zoho's increasing reach into its own market that it tried to buy the company. "They saw us as threatening because of the quality and price that we were delivering, and how quickly we were growing. I didn't sell because I built this company for a different reason than just to make money," Vembu shared with me. "Think about what we represent: we are recognized for creating great stuff, but we do it literally with a talent pool that none of our competitors would have ever hired."[38]

Zoho's international success does not come from producing software more cheaply than its competitors by paying low wages; it pays fair wages and provides great benefits for its employees. Zoho's competitiveness is a direct result of the way Vembu identifies and nurtures talent—and how that talent responds: by being fully engaged and extremely productive. By averagarian standards, Zoho shouldn't work; it's a company filled with employees who would never have been hired by most tech businesses and who are allowed to follow their own paths in the company, finding roles that allow them to contribute the most. Yet it does work. And Vembu is sure he knows why. "I have a strong math background and I know numbers. And I know that you are in big trouble if you start to think about individuals as numbers to be optimized on average," Vembu told me. "Treat individuals with respect, as individuals, and you will get out more than what you put in."[39]

FOSTERING INNOVATION AT MORNING STAR

Even in an industry such as manufacturing, where averagarianism has been the global standard for more than a century, a commitment to individuality can generate new and superior methods for getting things done. In fact, innovation is one of the biggest benefits of a commitment to individuality. Though Taylorist-modeled organizations like factories are often very good at managing costs and maximizing productivity within a set of constraints, they often have difficulty inspiring and harnessing creativity.

But even an industrial company can leverage the principles of individuality to create a culture that promotes individual initiative, nurtures individuality, and where innovative ideas are welcome regardless of their origin. The Morning Star Company has nurtured such a culture.

Founded in 1970 by Chris Rufer, Morning Star started out as a small one-truck, owner-operated company hauling tomatoes.[40]

Today, the Woodland, California, based company has over two hundred trucks, several factories, and thousands of employees. It controls 25 percent of tomato processing in California and produces 40 percent of tomato products consumed in the United States every year, making it the largest tomato-processing company in the world.[41] If you've had Campbell's tomato soup, Ragu spaghetti sauce, or Heinz ketchup, then odds are you've had Morning Star's products.[42]

On the surface, Morning Star's operations seem to be a perfect fit for a Taylorist model: a complex industrial process spanning fields and factories that churns hundreds of millions of tons of tomatoes each year with such efficiency it consistently has the lowest price in the industry.[43] But Frederick Taylor would likely have reeled with confusion if he knew what actually went on inside Chris Rufer's company.

At Morning Star, there are no managers. For that matter, there are no rigid titles, and virtually no hierarchy. Paul Green, a veteran of Morning Star who leads its training and development efforts, explained to me the philosophy behind such a radical business model: "All organizations are based upon fundamental assumptions about human beings, whether they know it or not. At Morning Star, we believe the individual is the single most important entity, and we do everything we can to promote the power of the individual."[44]

This is no bumper sticker cliché. At every level of its organization—or, more accurately, at every link in its organizational web—Morning Star is tacitly committed to the principles of individuality through what the company calls a "self-management" philosophy. Its system is set up to organically adapt to each employee's jaggedness, match employees to the contexts where they are effective, and empower individuals to pursue their own paths.[45] The focus on individual freedom and responsibility is best characterized by the personal mission statement.

Each employee drafts her own mission statement that defines how she will contribute to the company's overall mission and describes how she will achieve her goals and objectives. All employees who will be affected by her goals and activities must sign off on the statement. Employees are given enormous latitude to achieve their mission—including, for example, the right to make purchases to fulfill their mission—but are also held accountable by their peers (rather than a boss) for achieving or failing to achieve their goals and objectives.[46]

This is a very different way of thinking about job performance, and many new employees find it difficult to handle. Morning Star spent years trying to determine what kinds of personal qualities predicted success at the company, analyzing things like intelligence, personality, and education. They failed to find any meaningful correlations—except one: "People who have already worked a long time as managers at other companies can't figure out what to do," Paul Green told me. "They can't handle the freedom and the fact that they can't simply issue unilateral orders. But whenever people come here who don't know what it's like working anywhere else—or who didn't fit in at other places—they very quickly and naturally find a place for themselves."[47]

Like all Morning Star employees, Green does not have a title, though he is currently responsible for communicating the company's core principles throughout all of its divisions. He started out as a seasonal employee in 2006 maintaining a large industrial machine, known as a finisher, that whirls tomatoes around in a giant metallic cylinder to separate the skins from the fruit while losing as little tomato juice as possible. "This was a pretty dull job," Green told me. "But it was communicated to me from day one that at Morning Star I was free to modify my job however I wanted, provided that it furthered the mission of the company, and as long as I could convince any employees who would be affected by my changes that it was a good idea."[48]

Green wondered if a different setup to the finishing machine might separate tomato skins more effectively. He came up with an experiment: he ran a number of different finishers at different settings and recorded the results at fifteen-minute intervals for several months. Most companies would frown upon a newly hired seasonal maintenance worker setting up his own private engineering experiment on expensive equipment vital to the daily operation of the business—and that is an understatement. At most places, a factory temp is liable to be fired if he starts messing with the heart of the assembly line. But Green made his case to each stakeholder who would be affected by his proposed use of the finishers. "They were all very supportive," Green explained, "because I provided them with a very clear description of all the parameters of the experiment and what we stood to learn."[49]

After conducting his experiment, he discovered that there was indeed a different setting for the finishers that made them 25 percent more efficient, and Morning Star quickly adjusted all the machines to the new setting. He was soon hired full-time and has worked there ever since.

Green also told me the story of another colleague, someone we'll call Abe, who was initially hired as a manual laborer in the factory. Abe had always enjoyed tinkering, and while he was working for Morning Star he began playing around with the machines and equipment, even though this did not fit the duties he had been hired to perform. Gradually, he acquired a reputation as the go-to guy if you wanted some piece of machinery repaired or improved upon. So he made the case to his coworkers that he should be hired as a "master tinkerer," a position that had never existed at Morning Star before; and not only that—Abe asked for a budget to put together his own workshop where he could expand his range of tinkering.[50]

Morning Star does not automatically give blank checks to employees to pursue anything they fancy, like building their own work-shop. Many similar requests are often turned down. But Abe had demonstrated the tangible benefits of his tinkering to his cowork-ers, so they supported his proposal—along with a salary far higher than he would have received at any other company. Green told me, "Here's a guy who, based on his background and qualifications, you might easily have guessed would have never been much more than a laborer—let alone attain a career—but at Morning Star he found a role for himself where he is a recognized expert in the mechanics of factory machinery."[51]

That sense of belonging and personal purpose—that sense that you can add value to the company by sharing your ideas, and that the ideas will be listened to and, if they are good ideas, imple-mented—is at the heart of Morning Star's success. In traditional Taylorist organizations, that freedom to innovate is often curtailed or employees are discouraged to participate in the innovation process by the nature of their hierarchical structures. In such organizations, a specific division might be dedicated to innovation—research and development, for instance—or high-priced management consultants might be hired to suggest innovative processes or new products, but for most roles—like my job in the layaway department—ingenuity is actively frowned upon.

In contrast, innovation occurs frequently and organically in an individual-centric company like Morning Star, where temporary workers conduct assembly-line experiments and day laborers tin-ker with business-critical equipment. When you take individual-ity seriously—when you set up a business designed to embrace that individuality—innovation occurs everywhere, all the time, at every link of the network, because every employee is transformed into an independent agent tasked with figuring out the best way of doing her job and contributing to the company.

"We're not any kind of charity organization—every employee

must earn their place here," emphasizes Green. "But Morning Star gives everyone the freedom to earn their place. People are happiest when they have control over everything that's important to them."[52]

WIN-WIN CAPITALISM

Seventy years ago, Taylorism was considered "a characteristic of American civilization." But the principles of individuality show us the path to a better society, one that embraces individual freedom, initiative and responsibility, without sacrificing free enterprise. Costco, Zoho, and Morning Star demonstrate that when an organization makes the decision to value the individuality of its employees, it is not only the employees who win—the system wins, too, and wins bigger than ever. This is win-win capitalism, and it's available to any business in any industry in any country.

One final lesson can be learned from the success of Costco, Zoho, and Morning Star: if you make the decision to value individuals, that decision must be an unwavering commitment. These benefits— employee engagement, increased productivity, and widespread innovation—will not materialize if individuality is a fair-weather passion. "Even when people want to bet on this idea of investing in the individual, they get nervous when times get tough," Sinegal told me. "They close a factory, they lay off people, all to recoup a few cents here and there. Costco gave employees a raise during the recession because we knew they were struggling. It requires that you keep a clear eye on your purpose and not lose sight of that."[53] Zoho's Vembu came to a similar conclusion, telling me, "I want to hold on to an employee for life. It is a total commitment. But it's the commitment that makes a difference."[54]

I'm not saying that every company should attempt to replicate what Costco, Zoho, and Morning Star have done. Individuality demands that you think through what the principles mean for your particular business, and you build your business around that. But I am saying that it is possible for any business and any manager to implement the principles of individuality, and when you do—when you choose to invest in individuals—those individuals become loyal, driven, and passionate. It is possible to have engaged and productive employees who help companies win at the bottom line, even in the most averagarian of industries. You just can't have them on average.

CHAPTER 8

REPLACING THE AVERAGE IN HIGHER EDUCATION

When I started college in Ogden, Utah, I was desperate for a way out of a life of hardship and welfare. I needed a path to a better career that would allow me to provide for my wife and two sons, and this pathway needed to fit my incredibly tight financial constraints. Enrolling at Weber State University was the first step on this path, but nothing about my education came easy. During my first two years at college, I took all of my courses at night so I could continue working a full-time job during the day. Even so, my meager wages cooking bagels and selling electronics were never enough to cover all of my family's needs. Each month, without fail, we had to pick one bill not to pay. My wife sold as much blood plasma as she was legally allowed. I borrowed diapers from neighbors. We stole toilet paper from public restrooms.

My story is not so different from countless other families who endure hardship so that they, or their children, can graduate from college. The calculation behind these sacrifices is both rational and practical: we believe, correctly, that higher education is the single most important gateway to opportunity in our society. We are willing to do just about anything to obtain a diploma because we expect that it will give us or our children the best possible chance for a good job, a good income, a good neighborhood, a good life.

For anyone who looks at the value of a college degree in these pragmatic terms—and you can count me as one of them—the implicit purpose of higher education is to prepare students for their self-chosen careers at an affordable price. Perhaps you think higher education should have other goals, too, like promoting critical thinking, instilling an appreciation for the arts, or simply exposing students to new ideas. I agree that there are several other worthwhile goals that have a place in the mission statement—but I believe they must all be secondary to the primary goal of career preparation. In college, I learned critical thinking and social values and a lot of other wonderful things that made me a better person. But at the end of all those difficult years, if I had not obtained a good job that suited me, I would have considered the experience a failure.

If we agree with this practical goal for higher education, we cannot help but conclude that our current system is falling short.[1] Too many graduates cannot find a job in their field (31 percent, in one recent study by CareerBuilder);[2] too many employers cannot fill good-paying jobs (35 percent, according to the Manpower Group),[3] and too many employers report that the graduates they do hire are not equipped for their jobs.[4] And I doubt I have to try very hard to convince you that costs are out of control, but here is one telling fact: the cost of a college degree has risen 538 percent since 1985.[5] To put that into perspective, during that same period, medical costs rose by 286 percent.[6] Americans now have 1.1 *trillion* dollars in student loan debt,[7] more than all American credit card debt combined. I still owe a sizeable amount of money (enough to buy a very nice house in many parts of the U.S.) in student loans, a debt that hovers like a storm cloud over my financial future.

It is easy to imagine that it is the universities' fault we are all in this situation. It's not—or at least, no more than capitalism is at fault for some companies treating workers like statistics.

Like so much of the business world, the educational model of our system of higher education (and, just as important, its business model) is based on Taylorism.[8] Our contemporary universities are caretakers of an averagarian system they inherited that enforces the conviction that the system is more important than the individual and compels the standardization of all educational processes. The shortcomings of our system—its costs and, most important, the gap between what graduates learn and their ability to get a job—are due to a deeply entrenched averagarian architecture that was established long ago.

THE SAME, ONLY BETTER

Regardless of what colleges and universities *believe* their mission is today—whether it's to encourage problem solving and critical thinking or challenge students' viewpoints or some other worthy humanistic goal—our existing system of higher education was designed a century ago very explicitly to sort students by ranking them based on their performance in a standardized curriculum. High school students with the best grades and test scores go the best colleges, and then the college students with the best grades get the best jobs, as well as admission into the best professional schools. The system is the educational equivalent of the Norma look-alike competition, since its relentless focus on one-dimensional rankings compels every student to do exactly the same things the average student does. *Be the same as everyone else, only better.*

Even before they enter college, the system pressures students to conformity: if they want to get admitted to a good college, students need to take the same classes, tests, and extracurricular activities that everyone else is taking—but do better than everyone else. Once in college, students have to take the same courses as everyone in their major, in the same amount of time, to be ranked against the average, and to earn at the end of four years an undifferentiated diploma—all at a huge financial cost to them and their parents.

Judy Muir is a college admissions consultant based in Houston, and she understands this problem of conformity better than anyone.[9] She has dedicated her life to helping high school students get into college and succeed there, and for my money she's the best at what she does. She consults for the children of celebrities, presidents, and wealthy Europeans and Middle Easterners, though most of her clients are middle-class teens. She also does more than her share of pro bono consultations for underprivileged youngsters. Muir helps parents and teenagers make sense of the complex and daunting process of applying to college. But if you sit down with Muir, it doesn't take long before she vents her abiding frustration.

"The process is set up to ignore everything about the individuality of the student; it's all about average, average, average, select, select, select, leading teens to sublimate their own identity in the pursuit of the façade they think admissions officers want," Judy told me. "This is what the system has done to people, this runaway system that compares everyone against an average. Kids try to doctor their essay, they take internships they don't believe in. Overseas they cheat on their SATs. One of the most common questions I get is how many hours of community service do I need to do to get into this or that college. What I always tell them is that the only path to a life of excellence is by understanding and developing your own unique individuality. Instead, too many parents and kids focus on *hiding* their individuality instead of *developing* it, all because they are trying to stand out on the exact same things that everyone else is trying to stand out on."[10]

Bill Fitzsimmons, the Dean of Admissions and Financial Aid at Harvard, agrees, telling me, "Getting into college is usually a game of averages, except people are mortgaging their homes to play the game of averages. You're trading in your uniqueness to be like everyone else, in the hope that you can be a little bit better at the thing that everybody else is also trying to be. But if you're just playing the averages, then, on average, it won't work."[11]

So why are we all so willing to continue to play the game of averages when we know how flawed one-dimensional rankings of talent actually are? There is no scientific evidence that a sixteen-year-old's performance on a standardized test, or how many churches a seventeen-year-old helped build in Costa Rica, is meaningfully connected with becoming a Supreme Court justice or founding a successful start-up or discovering a cure for cancer. But as long as everyone else is playing the game of averages—and as long as universities and employers continue to play the game—there is a real cost for any student who chooses not to play.

So at every turn, students and their families make all kinds of sacrifices, taking on a staggering amount of debt, doing their best to conform themselves to a narrow and ruthless system based on a nineteenth-century notion of ranking—to receive a diploma that is no longer even a reliable guarantor of a job. The promise of our averagarian system of higher education keeps going down, while the costs imposed by the system keep going up.

If the *architecture* of higher education is based upon the false premise that students can be sorted by their rank—that a standardized, institution-centered system is necessary in order to efficiently separate the talented students from the untalented ones—then no matter how great the triumphs this system might produce, it is still guaranteed to produce some failures that we simply cannot tolerate as a society. Addressing these failures will require more than doubling down on the status quo: it will require committing to valuing the individual over the system, and changing the basic architecture of higher education so that the individual student truly comes first.

This might seem like an idea that sounds good in theory, but in practice is impossible to implement. But it turns out that the path to an individualized system of higher education, while not simple or easy, is reasonably straightforward, practical, and is already happening in colleges and universities around the world with great success.

To transform the averagarian architecture of our existing system into a system that values the individual student requires that we adopt these three key concepts:

- Grant credentials, not diplomas
- Replace grades with competency
- Let students determine their educational pathway

These concepts offer a blueprint for establishing an educational system that is consistent with the principles of individuality, and that will help *all* students choose and get trained for a career.

GRANT CREDENTIALS, NOT DIPLOMAS

Our current system of undergraduate education is standardized around one defining educational element—the four-year degree or diploma. For centuries the diploma and all the traditions surrounding its attainment—graduation ceremony, caps and gowns—have signaled to the community a student's achievement of a milestone, an educational rite of passage.

The problem is the requirements for a bachelor's degree are, to a large extent, arbitrary: no matter what subject you might pursue in college, the degree almost always requires the same four years. Whether you major in German literature or business administration or molecular biology—in each case, a bachelor's degree takes nearly the same total number of credit hours (what is known in the education field as "seat time") stretched over the same billable number of semesters.[12] It doesn't matter how difficult your chosen subject is, how fast or slow you learn, whether you go to a small private college or a sprawling public university, or whether you have mastered the necessary skills for your intended career: as long as you log the necessary hours of seat time (and do not fail a class), you will get a diploma. This, advocates of the four-year degree argue, results in a kind of "equality" of this rank across different fields.

Using the diploma as the basic unit of education introduces some obvious shortcomings into the system. If you finish all four years of seat time for a bachelor's degree in mechanical engineering and pass all of your courses—*except* for a single class in the humanities—you won't get a diploma. (You would still have to pay for four years of tuition, though.) It does not matter how well prepared you might be for a job as a mechanical engineer, if you do not complete every requirement set by the university, you don't get a diploma. Conversely, you could fulfill all the requirements for a computer science degree from an Ivy League college—and still not be equipped for a job as a computer programmer.[13]

There is a logical alternative to diplomas as the basic unit of educational achievement: *credentials*.[14] Credentialing is an approach to education that emphasizes awarding credit for the smallest meaningful unit of learning. For example, you might earn a credential for Java programming for websites, the history of World War I, pastry baking, or the climatology of Asia. Some credentials can be obtained after a few classes or even one class, whereas some may take a year or longer. Credentialing offers a more flexible and finer-grained level of certification of your skills, abilities, and knowledge.

Credentials can be combined ("stacked") to create more advanced credentials. For example, let's say you want to become a video game designer. Instead of pursuing a bachelor's degree in computer science, you might get credentials in programming theory, mobile device programming, computer animation, and graphic design. The completion of all four of these credentials might qualify you for a combined "mobile device-based video game design" credential. Similarly, if you want to be an astrophysicist who studies dark matter, you would proceed through a wide range of credentials in math, physics, astronomy, and research methods that would ultimately qualify you for a "dark matter astrophysics" credential. With credentialing there are no undergraduate programs that compel you to pay exorbitant tuition to a single university for four years to earn the necessary seat-hours for a standardized degree. Instead, you can pursue as few or as many credentials as you need to prepare for the career you want.

While the idea of credentials may seem a bit radical, the reality is that it has been an important part of skill-based education for a long time. For example, MIT already offers several credentialing programs (they call them "certificates"), including credentials in areas like supply chain management, managing complex technical projects, and big data (to name just a few).[15]

Virginia, meanwhile, has a large-scale state-sponsored program offering credentials in several industries, including information technology, cybersecurity, advanced manufacturing, energy, and health care.[16] The jobs that credentialed graduates obtain in these industries pay well and offer long-term career opportunities. The program requires approximately two to three weeks of full-time training in a simulated work environment and costs a total of $250 for each credential (the remaining costs are shouldered by industry, who are getting employees trained in the skills they need). So far, 93 percent of credentialed graduates from the program have obtained jobs. According to Governor Terry McAuliffe, the program has a goal of delivering nearly half a million credentials by 2030.[17]

There's nothing special about the particular fields targeted in the Virginia credentialing initiative—they were merely those fields with a known shortage of qualified job candidates—and there is no reason that credentialing cannot be extended to include everything taught in higher education, from French drama to quantum physics to cinematography.

Another recent educational development promises to make credentialing even more viable. Massively Open Online Courses, commonly referred to as MOOCs, are online courses offered by universities that do not require students to first be admitted into the university in order to enroll. Over the past decade, hundreds of universities have begun to offer MOOCs on every topic from Asian art to zoology. Much of the focus on MOOCs has been on their capacity to deliver online learning experiences at a discount or even for free. But I think the most innovative aspect of MOOCs is not their low cost or the fact that they are online, but rather the fact that many leading MOOC providers, including Harvard and MIT, have begun to offer credentials (such as certificates) for students who complete these courses.[18]

MOOCs point the way to what a fully developed individualized credentialing system might look like: no more undergraduate programs where you are compelled to pay exorbitant tuition to a single university for four years in order to earn the necessary seat-hours for a standardized degree. Instead, you pursue as many credentials as you need, at the cost you want, on your own terms, in order to pursue the career of your choice.

REPLACE GRADES WITH COMPETENCY

The second element of our averagarian system of higher education that must be changed is its basic method of evaluating performance: grades. Grades serve as a one-dimensional ranking of ability—grades supposedly represent how well we've mastered a subject and thus measure our ability within that field. They also serve as a marker of a student's progress along the standardized, fixed-pace pathway to a diploma.

There are two related problems with relying on grades for measuring performance. The first, and most important, is they are one-dimensional. The jaggedness principle, of course, tells us that any one-dimensional ranking cannot give an accurate picture of an individual's true ability, skill, or talent—or, as psychologist Thomas R. Guskey wrote in *Five Obstacles to Grading Reform,* "If someone proposed combining measures of height, weight, diet, and exercise into a single number or mark to represent a person's physical condition, we would consider it laughable. . . . Yet every day, teachers combine aspects of students' achievement, attitude, responsibility, effort, and behavior into a single grade that's recorded on a report card and no one questions it."[19]

The other problem posed by grades is that they require employers to perform a complex interpretation of what a particular graduate's diploma actually means. A transcript gives employers very little direct knowledge of a student's skills, abilities, or mastery of a topic. All they have to go on is the rank of a university and the graduate's GPA.

Fortunately, there is a straightforward solution to this problem: replace grades with a measure of *competency.* Instead of awarding grades for accumulating seat time in a course, completing all your homework on time, and acing your midterm, credentials would be given if, and only if, you demonstrate competency in the relevant skills, abilities, and knowledge needed for that particular credential. Although the nature of competency will differ from field to field, competency-based evaluation will have three essential features.

The first is rather obvious: it should be pass/incomplete—either you have demonstrated the competency or you have not. Second, competency evaluations must be institution-agnostic. This means you should be able to acquire the necessary competency for a credential in whatever way you like. You can still take a course—in most cases, this is probably the best option—but you will not get special credit just for completing the course, like you do right now under the current system. If you can acquire the competency online, on your own, or on the job, that's great—you do not need to pay for a course.

The third feature of competency-based evaluations of performance is that they should be professionally aligned. Obviously, that means professional organizations, as well as employers who will be hiring individuals with the credentials, should have some input into determining what constitutes competency for a particular profession-related credential. Of course, I am not saying employers should be the *only* ones to decide—that would be incredibly shortsighted—but I am saying they should have a genuine seat at the table. This will help ensure a tight, flexible, and real-time match between what students learn and what they will need to succeed in their jobs.

Does the idea of an industry-aligned, competency-based approach to education seem far-fetched? It is already here. Consider, for example, Western Governors University.[20] WGU is a nonprofit university that offers programs in business, information technology, health care, and teaching. Nineteen governors founded it in 1997 as an innovative strategy to better prepare students to work in particular high-need careers. The curriculum at WGU is entirely online, enabling students to move through material at their own pace. And though WGU grants degrees rather than credentials, students earn credit toward a degree by demonstrating competency, not by earning seat time in class. WGU also allows students to get credit for material they already know through competency exams without having to sit through an unnecessary course. Tuition at the school supports the notion of self-pacing: $6,000 covers as many courses as you can finish in two semesters of time.[21]

To ensure the industry-specific relevance of their programs, WGU has a two-step process for defining competency in a particular subject. The first is the "Program Councils"—panels of industry and academic experts who, together, define what a graduate in that area should know and be able to do to succeed at a job. The second is the "Assessment Councils," which consist of national experts who work to create competency exams that assess whether students have mastered the necessary material. Most important, WGU relies on industry-accepted assessments whenever possible rather than inventing its own.[22] Since WGU graduates have demonstrated competency in their field, they are attractive to employers.

WGU is not alone. More than two hundred schools are currently implementing or exploring competency-based forms of evaluating performance. There is even a consortium of universities working together to develop standards for scalable competency-based programs. Replacing grades with competency-based measures of performance will ensure that students can learn at their pace, and be judged according to their abilities.[23]

LET STUDENTS DETERMINE
THEIR EDUCATIONAL PATHWAYS

Granting credentials instead of degrees and replacing grades with competency-based evaluations are necessary for higher education to support individuality, but they are not sufficient. Today, universities control almost every aspect of your educational pathway. First and foremost, the university decides whether or not to admit you into one of its diploma-granting programs. If you do get admitted, the university dictates the requirements you must fulfill to obtain a diploma—and, of course, how much you pay for the privilege. Just about the only aspect of your education you *do* have control over is what university to apply to, and what to major in. We must cede more control to individual students by ensuring that our educational architecture supports *self-determined pathways*.

We can accomplish this by building on the competency-based credentialing foundation and focusing on two additional features of the higher education system. First, students should have *more* educational options to choose from than the ones offered by any single university. Second, the credentialing process should be independent of any particular institution, so that students have the ability to stack their credentials, no matter how or where they earned them.

In this system, students should be able to take a course anywhere: online or in a classroom, at an employer's training center or a local university. You could take a huge online course with thousands of students from all over the world, or get a local tutor to instruct you one-on-one, face-to-face. You could take an evening course once a week for six months, or an immersive two-week crash course. You could seek out high-intensity instructors who drive their students hard, or teachers who prefer to gently guide their students without pressing them. You could get all your credentials through courses from one institution, or stack together credentials from a variety of institutions. Or, in many cases, you could simply learn the material on your own, at your own pace, for free. The choice is up to you. Select the credentials pathway that helps *you* master the relevant knowledge, skills, and abilities, according to your own jagged profile, if-then signatures, and budget.

Self-determined pathways benefit students in many ways. Say you start out pursuing one stacked credential—maybe you are going after a neuroscience credential to one day become a researcher. You obtain a neuroanatomy credential and a neural systems credential, but discover that you like helping and interacting with people too much to spend your career focused on the physiological minutiae that is part of the daily grind of a bench scientist. So you decide to switch career goals and pursue a clinical psychology credential instead. The relevant neuroscience credentials that you have already obtained can be restacked and applied toward the clinical psychology credential. Or, if you decide that discussing people's problems does not quite suit you either, you could build on your existing credentials and restack them toward a career in marketing medical devices.

Right now, if you decided to switch majors in the middle of a traditional four-year neuroscience program, you would either need to shell out additional tuition as you make up for the classes you missed, or attempt to take an extra-large course load to finish in time, or maybe you would just complete the neuroscience degree and then apply to clinical psychology graduate programs or business schools—investing four years on a subject you are not fond of on your way to more years and more tuition for learning the subject you are really interested in.

With self-determined competency-based credentialing, there are fewer penalties for experimenting in order to discover what you are truly passionate about, and even fewer costs for switching horses midstream. In fact, if it is designed to support self-determination the entire educational system should encourage you to constantly re-assess what you like to do and what you might be good at, and give you a natural way to adjust your career plans as you go along according to what you learn about yourself and according to the changing job marketplace.

One of the most common reactions I hear from people when they first learn about self-determined educational pathways is, "So you are telling me that we expect college students to make their own decisions? Have you *met* today's college kids?" Though I won't disagree that a nineteen-year-old is more apt to make a foolish mistake than a forty-year-old, I am also skeptical of any system that tells us we cannot trust people to make decisions for themselves. Indeed, the notion that we should take away individuals' abilities to make decisions and allow the system to decide is quintessential Taylorism—the kind of thinking that got us into trouble in the first place.

That is the choice we are presented with: Do we want a system of higher education that compels each student to be like everyone else, only better? Or do we want a system that empowers each student to make her own choices?

EDUCATION IN THE AGE OF INDIVIDUALS

These three concepts—granting credentials, not diplomas; replacing grades with competency; and letting students determine their educational pathways—can help transform higher education from a system modeled after Taylorist factories that values top-down hierarchy and standardization, to a dynamic ecosystem where each student can pursue the education that suits her or him best.

A self-determined, competency-based credentialing system is also more closely aligned with the principles of individuality. It fulfills the jaggedness principle, since it allows students to figure out what they like, what they are good at, and what is the best way to pursue these interests. It fulfills the context principle by evaluating students' competency in a context as close as possible to the professional environment where they will actually perform. And it fulfills the pathways principle by allowing each student to learn at their own pace, and follow a sequence that is right for them.

Perhaps more important, adopting these concepts would help

solve the conformity problem: instead of trying to be like everyone else, only better, students will strive to be the very best version of themselves. Instead of playing the game of averages to get into a high-ranked university, you strive for professional excellence. Instead of competing with other students to be the best possible University Applicant, you compete with other students to be the best possible hire for an architectural firm, or an anthropology research laboratory, or a children's fashion designer. In this system, you get to be exactly who you are, not who the system tells you to be.

In addition, a self-determined, competency-based credentialing system would also put us on a path to solve the problem of endlessly rising educational costs. In an individualized system, you pay for exactly the credentials you want and need—and nothing more. Instead of one institution locking you into paying four years of tuition, different institutions will compete to offer you the best possible credential at the lowest possible price. Some institutions that adopt these elements might choose to emulate Western Governors University's "all you can learn" approach, where you pay a fixed fee and get all the training you wish from the institution. Other institutions might follow the lead of Arizona State University, which partnered with Harvard EdX to create a pioneering approach to online classes where first-year students only pay if they successfully complete them.[24]

An individualized educational system based on competency and credentialing also creates a much better match between students and employers, because the value and availability of credentials adjusts in real time according to the realities of the ever-changing job market. For example, if a new programming language starts to sweep through Silicon Valley, companies will quickly announce they are looking for individuals who are credentialed in the new language. Similarly, if the automobile industry switches away from an old engine style, there will be immediate pressure to reduce the engineering credentials that feature the outmoded technology. This provides students with tremendous flexibility to adjust their own pathways to take advantage of the changing market. Any student, at any moment, can see which credentials are valued by the companies they like, in geographic regions where they want to work, in industries where they want a career. They can compare costs, pathways, and difficulty of pursuing credentials and balance this with the potential salary and personal fit of various jobs.

At the same time, businesses and organizations can be assured of job applicants who have the skills and knowledge necessary for the job, because they can specify any combination of credentials that are needed for a particular job, no matter how demanding or complex, and because they have input into the competencies required for any given set of credentials. Employers can directly influence the pool of available employees, since they could offer to pay candidates to get a rare or unfamiliar credential, or even a new set of credentials.

It might seem like I'm saying that universities are the problem, or that universities are done for. No, I love universities. They provided me with the opportunity to attain a better life, and today one even pays part of my salary. Universities are essential for a vibrant, healthy democracy and a thriving economy. But the present architecture of our higher education system is based on a false premise: that we need a standardized system to efficiently separate the talented from the untalented. No matter how great the triumphs this present system might produce, its architecture is still guaranteed to produce some intolerable failures—so we must strive to change it.

Universities need to start asking tough questions about their education model. But if we really want to revolutionize the higher education system and move toward this new approach to education, then we need the help of the business world. Universities are unlikely to change unless employers demand something different. As long as employers continue to demand diplomas and degrees, there's little incentive for universities to change their system. This revolution in individualized education will only come once employers recognize how they will benefit from it and start to hire employees based on credentials, rather than diplomas, and based on employees' demonstrated competency rather than on grades.

An individualized approach to higher education is not easy, but it is possible. It's already happening in colleges and universities around the world. And it will benefit everyone—students, employers, even universities themselves. It all starts with one decision—to value the individual.

CHAPTER 9

REDEFINING OPPORTUNITY

In 2003, the U.S. Third Infantry Division was advancing toward the North Baghdad Bridge spanning the Tigris River when they unexpectedly stumbled on a nest of enemy soldiers who began to shower rocket-propelled grenades down onto the American forces. The infantry called for air support, and the air force sent in Captain K. Campbell, whose call sign was "Killer C." Despite the fierce nickname, Killer C was quite short for a pilot. In 1952, Campbell would never have fit into a cockpit designed for the average pilot, but in 2003 this undersized pilot was flying an A-10 Warthog, a ferocious beast of an aircraft created to wreak havoc upon ground forces.[1]

As Campbell unleashed the Warthog's firepower upon the Republican Guard, a huge explosion rocked the entire plane. "It felt and sounded like being in a car accident," Campbell told me.[2] A surface-to-air missile had shredded the rear of the plane, severely damaging the tail, fuselage, cowling, and horizontal stabilizers. Be assured, these are all essential parts of an airplane. All the hydraulic gauges flatlined as the controls lit up with flashing "EMERGENCY" lights. The flaming Warthog began diving straight down into the center of Baghdad—and when Campbell tried to pull up, the flight stick failed to respond.

Campbell glanced down at the ejection handle and for a moment considered ejecting and parachuting to safety. But this would mean allowing the monster jet to crash down through the streets of a crowded metropolis. Instead, Campbell flipped a switch that shifted the plane to manual piloting. Moving the flight stick into manual means using your own arm strength to pull heavy steel wires anchored to rudders and flaps. The closest analogy to manual piloting might be driving a car without power steering—except in this situation, it was more like driving a dump truck at two hundred miles an hour without power steering or rear tires, while missiles are being shot at you. Warthog pilots practice manual piloting once during their entire training, and never practice manual landing for the simple reason that it is too dangerous.[3]

In an attempt at making the perforated plane easier to control, Killer C jettisoned all the Warthog's weapons except for a counter-measures pod permanently affixed to the left wing of the plane. The suddenly asymmetric weight of the plane caused it to roll sharply to the left. "It was a heart-stopping moment," Campbell told me.[4] "I thought I was going to roll right into the ground." Picture the scene: a pint-size pilot trying to muscle a mechanical juggernaut out of a death spiral using the same hand-powered controls employed by the Wright Brothers . . . and succeeding.

Campbell regained control and flew out of Baghdad back to the American base in Kuwait where another tough decision had to be made: whether to attempt a manual landing. Manual piloting is incredibly difficult under the best of conditions. Manual landing is far harder. Campbell knew that manual landings of the Warthog had been attempted exactly three times before. The first time, the pilot died. The second time, the plane crashed and burned. The third time was a success, though the plane was not in as rough shape as Campbell's was.[5]

"It took me an hour to fly back to the base, so during that time I started getting comfortable with the controls," Campbell told me. "Not everyone agreed that I should have tried to land. But I had a lot of time to think about everything, the specific factors of that day, the clear weather conditions, good visibility, my comfort with the controls, an experienced wingman, the fact that I had been flying manual with my left arm to keep my right arm fresh for the landing. I'm the one in the seat and on that day I made the decision to land."[6]

Campbell did not crash and did not burn. Instead, a fellow pilot reported that Campbell "landed in manual more smoothly than I landed with hydraulics."[7] Campbell—now a colonel working in the Pentagon—received the Distinguished Flying Cross and a commendation from the South Carolina legislature.[8] But the acknowledgment that meant the most was scrawled on the back of a napkin: "Thanks for saving our ass that day," signed by a member of the Third Infantry.[9]

EQUAL FIT

I hope I have captured just how incredible a pilot Killer C is. But I would never be telling you this story if the U.S. Air Force still insisted that our pilots fit inside a cockpit designed for an average pilot: Colonel Kim N. Campbell, whose real call sign is Killer *Chick,* is five foot four and weighs 120 pounds[10]—and she could not be any further from someone's idea of an "average" pilot.

There is an important lesson here about the nature of opportunity. When the military adopted Lieutenant Gilbert Daniels's radical idea of creating adjustable cockpits that fit any person's body, nobody was talking about expanding the pool of pilot talent, let alone advocating for gender equity. They just wanted their existing pilots to perform better. The air force did not get Campbell because they designed a female-friendly plane, they got Campbell because they made a commitment to build planes designed to fit the jagged profile of individual pilots, whatever their jaggedness might be. "When I climb into the Warthog," Campbell said to me, "the seat needs to go to its maximum height and the pedals go all the way back—but it fits."[11]

This is the lesson of Kim Campbell: *fit creates opportunity.* If the environment is a bad match with our individuality—if we cannot reach the controls in the cockpit—our performance will always be artificially impaired. If we do get a good fit with our environment—whether that environment is a cockpit, a classroom, or a corner office—we will have the opportunity to show what we are truly capable of. This means that if we want equal opportunity for everyone, if we want a society where each one of us has the same chance to live up to our full potential, then we must create professional, educational, and social institutions that are responsive to individuality.

This is not how we usually think about equal opportunity. During the Age of Average we have defined opportunity as "equal access"—as ensuring that everyone has access to the *same* experiences.[12] Of course, equal access is undoubtedly preferable to older alternatives such as nepotism, cronyism, racism, misogyny, and classism. And there is no doubt that equal access has improved society immensely, creating a society that is more tolerant, respectful, and inclusive.[13] But equal access suffers from one major shortcoming: it aims to maximize individual opportunity *on average* by ensuring that everyone has access to the same standardized system, whether or not that system actually fits.

Imagine if the Air Force had passed a policy to allow all men and women the opportunity to become fighter pilots if they had the "right stuff," but continued to create cockpits designed for the average pilot. The Air Force would have rejected Kim Campbell, not because she lacked the talent to be a world-class pilot, but because she didn't fit inside an average cockpit. It would be difficult to argue that this is equal opportunity.

Equal access is an averagarian solution to an averagarian problem. For generations, people have been discriminated against because of gender, ethnicity, religion, sexual orientation, or socioeconomic class. Our response to such discrimination has been to try to balance the scales of opportunity—*on average*. If we see that the Average Man of one group is receiving different access to educational, professional, legal, and medical experiences than the Average Man of another group, then averagarian thinking suggests that the fair thing to do is to try to make those two Average Men as similar as possible. This was the right thing to do in the Age of Average—because it was the best we could do to address unfairness in a standardized world.

But now we know there is no such thing as an average person, and we can see the flaw in the equal access approach to opportunity: if there is no such thing as an average person, then there can never be equal opportunity on average. Only *equal fit* creates equal opportunity.[14]

Equal fit may seem like a novel idea, but it is ultimately the same view of opportunity expressed by Abraham Lincoln, when he declared that government's "leading object is to elevate the condition of men—to lift artificial weights from all shoulders, to clear the paths of laudable pursuit for all, to afford all an unfettered start and a fair chance, in the race of life."[15] Equal fit is an ideal that can bring our institutions into closer alignment with our values, and give each of us the chance to become the very best we can be, and to pursue a life of excellence, as we define it.

The good news is that we have it within our power, right now, to implement equal fit as a new foundation for equal opportunity in society. We no longer need to compel people to conform to the same inflexible standardized system, because we have the science and technology to build institutions that are responsive to individuality. But this transformation from an Age of Average to an Age of Individuals will not happen automatically. We must demand it.

If we are looking for the institution where implementing equal fit would have the biggest immediate impact on opportunity, the place to start is clear: public education. Despite the fact that "personalized learning" is the biggest buzzword in education today, and despite efforts of many organizations seeking change in the system, almost everything in traditional educational systems remains designed to ensure students receive the same exact standardized experience. Textbooks are designed to be "age appropriate," which means they are targeted toward the average student of a given age. Many assessments (including many so-called high-stakes tests) are age or grade normed, which means they are based around the average student of that age or grade.[16] We continue to enforce a curriculum that defines not only what students learn, but also how, when, at what pace, and in what order they learn it. In other words, whatever else we may say, traditional public education systems violate the principles of individuality.

Although it would not be easy, it's not difficult to imagine how to introduce equal fit into education. For starters, we can require that textbooks be designed "to the edges" rather than to the average; we can require that curricular materials be adaptive to individual ability and pacing rather than fixed based on grade or age; we can require that educational assessments be built to measure *individual* learning and development rather than simply ranking students against one another. Finally, we can encourage local experimentation and sharing of successes and failures to accelerate discovery and adoption of cost-effective, scalable ways to implement student-driven, self-paced, multipathway educational experiences.

We can also apply the principle of equal fit to social policies that influence the workplace, such as policies that influence hiring, firing, and pay. Imagine the talent that we can unleash by redesigning our schools and jobs to fit the individual, instead of fitting the averagarian system—even if that averagarian system is motivated by the best of intentions. We would unleash a society of Kim Campbells—a society of individual excellence.

RESTORING THE DREAM

James Truslow Adams coined the phrase "American Dream" in his 1931 book *The Epic of America,* which he published in the depths of the Great Depression. Adams argued for a view of the American dream that ran counter to the materialism of his time: "It is not a dream of motor cars and high wages merely, but a dream of social order in which each man and each woman shall be able to attain to the fullest stature of which they are innately capable, and be recognized by others for what they are, regardless of the fortuitous circumstances of birth or position."[17]

The original formulation of the American dream was not about becoming rich or famous; it was about having the opportunity to live your life to its fullest potential, and being appreciated for who you are as an individual, not because of your type or rank. Though America was one of the first places where this was a possibility for many of its citizens, the dream is not limited to any one country or peoples; it is a universal dream that we all share. And this dream has been corrupted by averagarianism.

Adams originally coined his term in direct response to the growing influence of Taylorism and the efficiency movement, which valued the system, but had "no regard for the individuals to whom alone any system could mean anything."[18] For Adams, the Taylorist view of the world was not only altering the fabric of society, it was altering the way people viewed themselves and one another, the way they determined their priorities, the way they defined the meaning of success. As averagarianism reshaped the educational system and workplace, the American dream came to signify less about personal fulfillment and more about the notion that even the lowliest of citizens could climb to the topmost rungs of the economic ladder.

It is easy to see why this shift in values occurred, and it is not nearly as straightforward as simple materialism. We all feel the weight of the one-dimensional thinking that has become so pervasive in our averagarian culture: a standardized educational system that ceaselessly sorts and ranks us; a workplace that hires us based on these educational rankings, then frequently imposes new rankings at every annual performance review; a society that doles out rewards, esteem, and adoration according to our professional ranking. When we look up at these artificial, arbitrary, and meaningless rungs that we are expected to climb, we worry that we might not fully ascend them, that we will be denied those opportunities that are only afforded to those who muscle their way up the one-dimensional ladder.

We worry that if we, or our children, are labeled "different" we will have no chance of succeeding in school and will be destined to a life on the lower rungs. We worry that if we do not attend a top-tier school and earn a high GPA, the employers we want to work for may not even look at us. We worry that if we answer a personality test in the wrong way, we may not get the job we want. We live in a world that demands we be the same as everyone else, only better, and reduces the American dream to a narrow yearning to be *relatively* better than the people around us, rather than the best version of ourselves.

The principles of individuality present a way to restore the meaning of the American dream—and, even better, the chance for everyone to attain it. If we overcome the barriers of one-dimensional thinking, essentialist thinking, and normative thinking, if we demand that social institutions value individuality over the average, then not only will we have greater individual opportunity, we will change the way we think about success—not in terms of our deviation from average, but on the terms we set for ourselves.

We are not talking about a future utopia; we are talking about a practical reality that is already happening all around us today. Our health-care system is moving toward personalized medicine, with the goal of equal fit for every patient. Competency-based credentialing is being tried out—successfully—at leading universities. Context-based hiring is already here and being spearheaded by pioneers like Lou Adler. Enterprises that have committed themselves to valuing the individual are achieving global success, like Costco, Zoho, and Morning Star. These are the places that provide us with a glimpse of what equal fit will actually look like. It's time for *all* institutions to embrace individuality and adopt equal fit as the necessary credo to restore the dream.

The ideal that we call the American dream is one we all share—the dream of becoming the best we can be, on our own terms, of living a life of excellence, as we define it. It's a dream worth striving for. And while it will be difficult to achieve, it has never been closer to becoming a reality than it is right now. We no longer need to be limited by the constraints imposed on us by the Age of Average. We can break free of the tyranny of averagarianism by choosing to value individuality over conformity to the system. We have a bright future before us, and it begins where the average ends.

ACKNOWLEDGMENTS

Putting together *The End of Average* was one of the most demanding adventures of my life, but fortunately I didn't make the journey alone. My partner on this trek was my colleague, my friend, my coauthor, Dr. Ogi Ogas. Ogi's blood and sweat courses through the pages of this book as much as my own, and I am proud to say the work that you hold in your hands is the result of our singular collaboration.

I am grateful beyond words for my incredibly talented editor at HarperOne, Genoveva Llosa. Her commitment to the ideas, and her passion for making the science relevant and actionable, made her an invaluable partner. The energy and engagement she devoted to the too-many drafts of the manuscript are an endless source of gratitude and professional respect. This book would never be what it is today without her vision, dedication, and guidance. Thanks, as well, to the stellar team at HarperOne: Hannah Rivera, Kim Dayman, Suzanne Wickham, and Lisa Zuniga.

I also want to extend a special thanks to Howard Yoon, my brilliant literary agent. He helped me shape a raw, exuberant idea into a commercial project, and contributed in many important ways to the final product.

This book would never have been possible without the boundless generosity of my colleagues at The Center for Individual Opportunity: Dewey Rosetti, Bill Rosetti, Debbie Newhouse, Parisa Rouhani, Walter Haas, and Brian Daly. Put simply, I would never have had the courage and foresight to pursue the big ideas in this book without their support, and each one of them has contributed in a significant way to *The End of Average*.

There is no single person to whom I am more indebted for my intellectual development than Kurt Fischer, who took me under his wing and taught me how to be a scientist and scholar. There has been no greater honor in my academic life than succeeding him as director of the Mind, Brain, and Education program that he established.

When it comes to the people whose ideas form the lifeblood of *The End of Average*, Peter Molenaar deserves special mention. His work—and life—is an inspiration to everyone who has ever felt that there is something fundamentally wrong with the way the world sizes you up. I am grateful for the hours of conversation, feedback, and support that he has given me, and for his tireless work on behalf of this new science.

In addition, several other scholars have influenced my thinking about individuality, most notably Jim Lamiell, as well as Lars Bergman, Anne Bogat, Peter Borkenau, Denny Borsboom, Alexander von Eye, Emilio Ferrer, Howard Gardner, Paul van Geert, James Grice, Ellen Hamaker, Michael Hunter, Michelle Lampl, Han van der Maas, David Magnusson, Mike Miller, Walter Mischel, John Nesselroade, Fritz Ostendorf, Yuichi Shoda, Robert Siegler, Esther Thelen, Jaan Valsiner, Beatrix Vereijken, and Jamil Zaki.

My earnest gratitude is owed to Karen Adolph, Lou Adler, Juliet Agranoff, Kelly Bryant, Colonel Kim Campbell, Todd Carlisle, Gilbert Daniels, Callum Negus-Fancey, Bill Fitzsimmons, Ashley Goodall, Paul Green, Mike Miller, Judy Muir, Yuichi Shoda, Jim Sinegal, and Sridhar Vembu for allowing me to interview them, and for sharing their ideas and insights with me.

Though I did not end up using nearly as much of their material as I originally expected, I would be remiss if I did not thank Paul Beale of the University of Colorado and Thomas Greytak of MIT, two world-class physicists who took the time to clarify in illuminating detail the nature of statistical physics to me, as well as the quantum mechanics of gases. I also want to thank Kevin Donnelly for helpful conversations about Adolphe Quetelet.

I would also like to personally mention Stacy Parker-Fisher, for her vision and passion; Ellice Sperber, the Oak Foundation, Wasserman Foundation, and the Walter & Elise Haas Fund for supporting my efforts to bring the ideas of individuality to a wider audience; Sandy Otellini for her strategic guidance; and Debbie Johnson for giving me the opportunity to present my ideas to the public for the first time at TEDxSonomaCounty.

Special thanks to Katie Zannechia, who has been a marvel at online marketing, social media, and strategy—smart, clever, enthusiastic, and always a step ahead of me; Mike Dicks, an incredibly gifted designer; Noah Gallagher Shannon for an incredibly thorough job as my primary fact-checker; and Tofool Alghanem for providing round after round of feedback on the ever-changing manuscript.

I would also like to thank David Sarokin, J.D. Umiat, and Bobbie Sevens, the ultra-competent researchers at Uclue.com; the United States Air Force for granting me permission to interview Colonel Campbell, and Del Christman for introducing me to the colonel; and the Cleveland Museum for granting me permission to use the image of *Norma*.

Chris Betke—thank you for providing the kind of services nobody ever wants, yet always ends up being essential, and for doing so with such panache; thank you, too, to Andrew Ferguson and Matthew Lynch, Jr.

One of the most essential and practical ways anybody can help a book improve and develop is by taking the time to read it and share their frank reaction, so I would like to thank the many wonderful people who critiqued drafts of the book: Debbie Newhouse, Parisa Rouhani, Reem AlGhanem, Basic AlGhanem, John and Sandy Ogas, Priyanka Rai, Chaitanya Sai, Elizabeth Ricker, Marianne Brandon, Amiel Bowers, Hama Gheddaf Dam, Kit Maloney, Deepti Rao, Chris Betke, Kalim Saliba, and Anna Sproul-Latimer and Dara Kaye of Ross-Yoon.

A big thank you to my parents, Larry and Lyda Rose, to whom I owe my enduring optimism and commitment to individuality. My father taught me one of the greatest lessons I've learned in life: that it is possible to respect and cherish tradition, while still questioning absolutely everything.

Finally, the greatest unsung heroes of *The End of Average* are the people who had to endure every absence, conversation, and deadline: My wife, Kaylin, and my sons, Austin and Nathan. There is not enough gratitude in the world to thank you for putting up with me. Without you, none of this would be possible.

NOTES

INTRODUCTION: THE LOOK-ALIKE COMPETITION

1. "USAF Aircraft Accidents, February 1950," Accident-Report.com, http://www.accident-report.com/Yearly/1950/5002.html.

2. Francis E. Randall et al., *Human Body Size in Military Aircraft and Personal Equipment* (Army Air Forces Air Materiel Command, Wright Field, Ohio, 1946), 5.

3. United States Air Force, *Anthropometry of Flying Personnel* by H. T. Hertzberg et al., WADC-TR–52–321 (Dayton: Wright-Patterson AFB, 1954).

4. Gilbert S. Daniels, interviewed by Todd Rose, May 14, 2014.

5. For an overview of this particular approach to typing, see W. H. Sheldon et al., *Atlas of Man* (New York: Gramercy Publishing Company, 1954).

6. Earnest Albert Hooton, *Crime and the Man* (Cambridge: Harvard University Press, 1939), 130.

7. Gilbert S. Daniels, "A Study of Hand Form in 250 Harvard Men" (unpublished thesis submitted for honors in the Department of Anthropology, Harvard University, 1948).

8. Daniels, interview.

9. Gilbert S. Daniels, *The "Average Man"?* TN-WCRD–53–7 (Dayton:

197

Wright-Patterson AFB, Air Force Aerospace Medical Research Lab, 1952).

10. Daniels, *The "Average Man"?*, 3.

11. Josephine Robertson, "Are You Norma, Typical Woman? Search to Reward Ohio Winners," *Cleveland Plain Dealer*, September 9, 1945.

12. Anna G. Creadick, *Perfectly Average: The Pursuit of Normality in Postwar America* (Amherst: University of Massachusetts Press, 2010). Note: The sculptures are available at Harvard Countway Library; "CLINIC: But Am I Normal?" *Remedia*, November 5, 2012, http://remedianetwork.net/2012/11/05/clinic-but-am-i-normal/; Harry L. Shapiro, "A Portrait of the American People," *Natural History* 54 (1945): 248, 252.

13. Dahlia S. Cambers, "The Law of Averages 1: Normman and Norma," *Cabinet*, Issue 15, Fall 2004, http://www.cabinetmagazine.org/issues/15/cambers.php; and Creadick, *Perfectly Average*.

14. Bruno Gebhard, "The Birth Models: R. L. Dickinson's Monument," *Journal of Social Hygiene* 37 (April 1951), 169–174.

15. Gebhard, "The Birth Models."

16. Josephine Robertson, "High Schools Show Norma New Way to Physical Fitness," *Cleveland Plain Dealer*, September 18, 1945, A1.

17. Josephine Robertson, "Are You Norma, Typical Woman? Search to Reward Ohio Winners," *Cleveland Plain Dealer*, September 9, 1945, A8; Josephine Robertson, "Norma Is Appealing Model in Opinion of City's Artists," *Cleveland Plain Dealer*, September 15, 1945, A1; Josephine Robertson, "Norma Wants Her Posture to Be Perfect," *Cleveland Plain Dealer*, September 13, 1945, A1; Josephine Robertson, "High Schools Show Norma New Way to Physical Fitness," *Cleveland Plain Dealer*, September 18, 1945, A1; Josephine Robertson, "Dr. Clausen Finds Norma Devout, but Still Glamorous," *Cleveland Plain Dealer*, September 24, 1945, A3; "The shape we're in." *TIME*, June 18, 1945; Creadick, *Perfectly Average*, 31–35.

18. Josephine Robertson, "Theater Cashier, 23, Wins Title of Norma, Besting 3,863 Entries," *Cleveland Plain Dealer*, September 23, 1945, A1.

19. Robertson, "Theater Cashier," A1.

20. Robertson, "Theater Cashier," A1.

21. Daniels, *The "Average Man"?*, 1.

22. Daniels, *The "Average Man"?*

23. Daniels, interview.

24. Kenneth W. Kennedy, *International anthropometric variability and its effects on aircraft cockpit design*. No. AMRL-TR-72-45. (Air Force Aerospace medical research lab, Wright-Patterson AFB OH, 1976); for an example of manufacturers implementing the design standards, see Douglas Aircraft Company, El Segundo, California, Service Information Summary, Sept.–Oct., 1959.

25. E. C. Gifford, *Compilation of Anthropometric Measures of US Navy Pilots*, NAMC-ACEL–437 (Philadelphia: U.S. Department of the Navy, Air Crew Equipment Laboratory, 1960).

26. L. Todd Rose et al., "The Science of the Individual," *Mind, Brain, and Education* 7, no. 3 (2013): 152–158. See also James T. Lamiell, *Beyond Individual and Group Differences: Human Individuality, Scientific Psychology, and William Stern's Critical Personalism* (Thousand Oaks: Sage Publications, 2003).

27. "Miasma Theory," *Wikipedia*, June 27, 2015, https://en.wikipedia.org/wiki/Miasma_theory.

28. "Infectious Disease Timeline: Louis Pasteur and the Germ Theory of Disease," *ABPI*, http://www.abpischools.org.uk/page/modules/infectious diseases_timeline/timeline4.cfm.

CHAPTER 1: THE INVENTION OF THE AVERAGE

1. Michael B. Miller et al., "Extensive Individual Differences in Brain Activations Associated with Episodic Retrieval Are Reliable Over Time," *Journal of Cognitive Neuroscience* 14, no. 8 (2002): 1200–1214.

2. K. J. Friston et al., "How Many Subjects Constitute a Study?" *Neuroimage* 10 (1999): 1–5.

3. Michael Miller, interviewed by Todd Rose, September 23, 2014.

4. Miller, interview.

5. L. Cahill et al., "Amygdala Activity at Encoding Correlated with Long-Term, Free Recall of Emotional Information," *Proceedings of the National Academy of Sciences, U.S.A.* 93 (1996): 8016–8021; I. Klein et al., "Transient Activity in the Human Calcarine Cortex During Visual-Mental Imagery: An Event-Related fMRI Study," *Journal of Cognitive Neuroscience* 12 (2000): 15–23; S. M. Kosslyn et al., "Individual Differences in Cerebral Blood Flow in Area 17 Predict the Time to Evaluate Visualized Letters," *Journal of Cognitive Neuroscience* 8 (1996): 78–82; D. McGonigle et al., "Variability in fMRI: An Examination of Intersession

Differences," *Neuroimage* 11 (2000): 708–734; S. Mueller et al., "Individual Variability in Functional Connectivity Architecture of the Human Brain," *Neuron* 77, no. 3 (2013): 586–595; L. Nyberg et al., "PET Studies of Encoding and Retrieval: The HERA model," *Psychonomic Bulletin and Review* 3 (1996): 135–148; C. A. Seger et al., "Hemispheric Asymmetries and Individual Differences in Visual Concept Learning as Measured by Functional MRI," *Neuropsychologia* 38 (2000): 1316–1324; J. D. Watson et al., "Area V5 of the Human Brain: Evidence from a Combined Study Using Positron Emission Tomography and Magnetic Resonance Imaging," *Cerebral Cortex* 3 (1993): 79–94. Also note, there is even known individuality in the hemodynamic response. See G. K. Aguirre et al., "The Variability of Human, BOLD Hemodynamic Responses," *Neuroimage* 8 (1998): 360–369.

6. Miller, interview, 2014.

7. Miller, interview, 2014.

8. His full name was Lambert Adolphe Jacques Quetelet. For biographical and background information, see Alain Desrosières, *The Politics of Large Numbers: A History of Statistical Reasoning* (Cambridge: Harvard University Press, 1998), chap. 3; K. P. Donnelly, *Adolphe Quetelet, Social Physics and the Average Men of Science, 1796–1874* (London: Pickering & Chatto, 2015); Gerd Gigerenzer et al., *The Empire of Chance: How Probability Changed Science and Everyday Life* (Cambridge: Cambridge University Press, 1989); Ian Hacking, *The Emergence of Probability: A Philosophical Study of Early Ideas about Probability, Induction and Statistical Inference* (Cambridge: Cambridge University Press, 1975); Ian Hacking, *The Taming of Chance* (Cambridge: Cambridge University Press, 1990); T. M. Porter, *The Rise of Statistical Thinking, 1820–1900* (Princeton: Princeton University Press, 1986); Stephen M. Stigler, *The History of Statistics: The Measurement of Uncertainty before 1900* (Cambridge: Harvard University Press, 1986); Stephen M. Stigler, *Statistics on the Table: The History of Statistical Concepts and Methods* (Cambridge: Harvard University Press, 2002).

9. Stigler, *History of Statistics,* 162.

10. Porter, *Rise of Statistical Thinking,* 47.

11. Porter, *Rise of Statistical Thinking,* 47–48.

12. T. M. Porter, "The Mathematics of Society: Variation and Error in Quetelet's Statistics," *British Journal for the History of Science* 18, no. 1 (1985): 51–69, citing Quetelet, "Memoire sur les lois des naissances et de la mortalite a Bruxelles," *NMB* 3 (1826): 493–512.

13. Porter, *Rise of Statistical Thinking,* 104.

14. I. Hacking, "Biopower and the Avalanche of Printed Numbers," *Humanities in Society* 5 (1982): 279–295.

15. C. Camic and Y. Xie, "The Statistical Turn in American Social Science: Columbia University, 1890 to 1915," *American Sociological Review* 59, no. 5 (1994): 773–805; and I. Hacking, "Nineteenth Century Cracks in the Concept of Determinism," *Journal of the History of Ideas* 44, no. 3 (1983): 455–475.

16. Porter, *Rise of Statistical Thinking*, 95.

17. S. Stahl, "The Evolution of the Normal Distribution," *Mathematics Magazine* 79 (2006): 96–113.

18. O. B. Sheynin, "On the Mathematical Treatment of Astronomical Observations," *Archives for the History of Exact Sciences* 11, no. 2/3 (1973): 97–126.

19. Adolphe Quetelet, "Sur l'appréciation des documents statistiques, et en particulier sur l'application des moyens," *Bulletin de la Commission Centrale de la Statistique (of Belgium)* 2 (1844): 258; A. Quetelet, *Lettres à S. A. R. Le Duc Régnant de Saxe Cobourg et Gotha, sur la théorie des probabilités, appliquée aux sciences morales et politique* (Brussels: Hayez, 1846), letters 19–21. The original data are from the *Edinburgh Medical and Surgical Journal* 13 (1817): 260–264.

20. T. Simpson, "A Letter to the Right Honourable George Macclesfield, President of the Royal Society, on the Advantage of Taking the Mean, of a Number of Observations, in Practical Astronomy," *Philosophical Transactions* 49 (1756): 82–93.

21. Stahl, "Evolution of the Normal Distribution," 96–113; and Camic and Xie, "Statistical Turn," 773–805.

22. Quetelet, *Lettres*, Letters 19–21.

23. Quetelet, *Lettres*, Letter 20.

24. Quetelet, *Lettres*, Letters 90–93.

25. Adolphe Quetelet, *Sur l'homme et le développement de ses facultés, ou Essai de physique sociale* (Paris: Bachelier, 1835); trans. *A Treatise on Man and the Development of his Faculties* (Edinburgh: William and Robert Chambers, 1842), chap. 1. A revised version of this book changed the title: *Physique sociale ou essai sur le développement des facultés de l'homme* (Brussels: C. Muquardt, 1869).

26. Stigler, *History of Statistics*, 171; quoting passage at page 276 of Quetelet, *Sur L'homme* (1835).

27. Quetelet, *Treatise*, 99.

28. Quetelet, *Treatise*, 276.

29. Hacking, "Nineteenth Century Cracks," 455–475; Kaat Louckx and Raf Vanderstraeten, "State-istics and Statistics, 532; N. Rose, "Governing by Numbers: Figuring Out Democracy," *Accounting* 16, no. 7 (1991): 673–692; and "Quetelet, Adolphe." *International Encyclopedia of the Social Sciences,* 1968; *Encyclopedia.com.* (August 10, 2015). http://www .encyclopedia.com/doc/1G2-3045001026.html.

30. John S. Haller, "Civil War Anthropometry: The Making of a Racial Ideology," *Civil War History* 16, no. 4 (1970): 309–324. The original report references Quetelet: J. H. Baxter, *Statistics, Medical and Anthropological, of the Provost Marshal-General's Bureau, Derived from Records of the Examination for Military Service in the Armies of the United States During the Late War of the Rebellion, of Over a Million Recruits, Drafted Men, Substitutes, and Enrolled Men* (Washington: U.S. Government Printing Office, 1875), 17–19, 36, 43, 52. Quetelet uses this result as proof of types (Quetelet, *Anthropometrie* [Brussels: C. Muquardt, 1871], 16); Quetelet, "Sur les proportions de la race noire," *Bulletin de l'acadimie royale des sciences et belles-lettres de Belgique* 21, no. 1 (1854): 96–100).

31. Porter, "Mathematics of society," 51–69.

32. A. Quetelet, *Du systeme et des lois qui social régissent him* (Paris: Guillaumin, 1848), 88–107, 345–346.

33. Mervyn Stone, "The Owl and the Nightingale: The Quetelet/Nightingale Nexus," *Chance* 24, no. 4 (2011): 30–34; Piers Beirne, *Inventing Criminology* (Albany: SUNY Press, 1993), 65; Wilhelm Wundt, *Theorie Der Sinneswahrnehmung* (Leipzig: Winter'sche, 1862), xxv; J. C. Maxwell, "Illustrations of the Dynamical Theory of Gases," *Philosophical Magazine* 19 (1860): 19–32. Reprinted in *The Scientific Papers of James Clerk Maxwell* (Cambridge: Cambridge University Press, 1890; New York: Dover, 1952, and Courier Corporation, 2013).

34. For biographical and background information on Galton see F. Galton, *Memories of My Life* (London: Methuen, 1908); K. Pearson, *The Life, Letters and Labours of Francis Galton* (London: Cambridge, University Press, 1914); D. W. Forrest, *Francis Galton: The Life and Work of a Victorian Genius* (New York: Taplinger, 1974); and R. E. Fancher, "The Measurement of Mind: Francis Galton and the Psychology of Individual Differences," in *Pioneers of Psychology* (New York: Norton, 1979), 250–294.

35. Jeffrey Auerbach, *The Great Exhibition of 1851* (New Haven: Yale University Press, 1999), 122–123.

36. Gerald Sweeney, "Fighting for the Good Cause," *American Philosophical Society* 91, no. 2 (2001): i–136.

37. Sweeney, "Fighting for the Good Cause." For information on changes in voting rights, see Joseph Hendershot Park, *The English Reform Bill of 1867* (New York: Columbia University, 1920).

38. Francis Galton, *Hereditary Genius: An Inquiry into Its Laws and Consequences* (New York: Horizon Press, 1869), 26. See the appendix for a discussion of some of the mathematical aspects of the "average man."

39. Sweeney, "Fighting for the Good Cause," 35–49.

40. Francis Galton, "Eugenics: Its Definition, Scope, and Aims," *American Journal of Sociology* 10, no. 1 (1904): 1–25.

41. Michael Bulmer, *Francis Galton* (Baltimore: JHU Press, 2004), 175.

42. Francis Galton, "Statistics by Intercomparison, with Remarks on the Law of Frequency of Error," *Philosophical Magazine* 49 (1875): 33–46.

43. Francis Galton, *Inquiries into Human Faculty and Its Development* (London: Macmillan, 1883), 35–36.

44. Francis Galton, *Essays in Eugenics* (London: The Eugenics Education Society, 1909), 66.

45. Piers Beirne, "Adolphe Quetelet and the Origins of Positivist Criminology," *American Journal of Sociology* 92, no. 5 (1987): 1140–69; for a broader treatment of the topic, see Porter, *Rise of Statistical Thinking*.

46. Quetelet, *Sur l'homme*, 12.

47. K. Pearson, "The Spirit of Biometrika," *Biometrika* 1, no. 1 (1901): 3–6.

48. William Cyples, "Morality of the Doctrine of Averages," *Cornhill Magazine* (1864): 218–224.

49. Claude Bernard, *Principes de médecine expérimentale*, L. Delhoume, ed. (Paris, 1947), 67, quoted in T. M. Porter, *The Rise of Statistical Thinking, 1820–1900* (Princeton: Princeton University Press, 1986), 160.

50. Claude Bernard, *An Introduction to the Study of Experimental Medicine* (New York: Dover, 1865; 1957), 138.

51. Joseph Carroll, "Americans Satisfied with Number of Friends, Closeness of Friendships," Gallup.com, March 5, 2004, http://www.gallup.com /poll/10891/americans-satisfied-number-friends-closeness-friendships .aspx; "Average Woman Will Kiss 15 Men and Be Heartbroken Twice Before Meeting 'The One', Study Reveals," *The Telegraph*, January 1, 2014, http://www.telegraph.co.uk/news/picturegalleries/howabout

that/10545810/Average-woman-will-kiss–15-men-and-be-heartbroken -twice-before-meeting-The-One-study-reveals.html; "Finances Causing Rifts for American Couples," AICPA, May 4, 2012, http://www.aicpa .org/press/pressreleases/2012/pages/finances-causing-rifts-for-american -couples.aspx.

CHAPTER 2: HOW OUR WORLD BECAME STANDARDIZED

1. J. Rifkin, *Time Wars: The Primary Conflict in Human History* (New York: Henry Holt & Co., 1987), 106.

2. For biographical information on Taylor: see Robert Kanigel, *The One Best Way: Frederick Winslow Taylor and the Enigma of Efficiency* (Cambridge: MIT Press Books, 2005).

3. Charles Hirschman and Elizabeth Mogford, "Immigration and the American Industrial Revolution from 1880 to 1920," *Social Science Research* 38, no. 1 (2009): 897–920.

4. Kanigel, *One Best Way*, 188.

5. Eric L. Davin, *Crucible of Freedom: Workers' Democracy in the Industrial Heartland, 1914–1960* (New York: Lexington Books, 2012), 39; Daniel Nelson, *Managers and Workers* (Madison: University of Wisconsin Press, 1995), 3; and J. Mokyr, "The Second Industrial Revolution, 1870–1914," August 1998, http://faculty.wcas.northwestern.edu/~jmokyr /castronovo.pdf.

6. Frederick Winslow Taylor, *The Principles of Scientific Management* (New York: Harper & Brothers, 1911), 5–6.

7. Taylor, *Principles of Scientific Management*, 7.

8. Taylor Society, *Scientific Management in American Industry* (New York: Harper & Brothers, 1929), 28.

9. Taylor, *Principles of Scientific Management*, 83.

10. Kanigel, *One Best Way*, 215.

11. Hearings Before Special Committee of the House of Representatives to Investigate the Taylor and Other Systems of Shop Management Under Authority of House Resolution 90, no. III, 1377–1508. Reprinted in *Scientific Management*, Frederick Winslow Taylor (Westport: Greenwood Press, 1972), 107–111.

12. Taylor, *Principles of Scientific Management*, 25.

13. Frederick W. Taylor, "Why the Race Is Not Always to the Swift," *American Magazine* 85, no. 4 (1918): 42–44.

14. Maarten Derksen, "Turning Men into Machines? Scientific Management, Industrial Psychology, and the Human Factor," *Journal of the History of the Behavioral Sciences* 50, no. 2 (2014): 148–165.

15. Taylor, *Principles of Scientific Management*, 36.

16. Kanigel, *One Best Way*, 204.

17. From a lecture on June 4, 1906 (cited in Kanigel, *One Best Way*, 169).

18. Frederick W. Taylor, "Not for the Genius—But for the Average Man: A Personal Message," *American Magazine* 85, no. 3 (1918): 16–18.

19. Taylor, *Principles of Scientific Management*.

20. Thomas K. McCraw, *Creating Modern Capitalism: How Entrepreneurs, Companies, and Countries Triumphed in Three Industrial Revolutions* (Cambridge, MA: Harvard University Press, 1997), 338; http://www.newyorker.com/magazine/2009/10/12/not-so-fast; and Peter Davis, *Managing the Cooperative Difference: A Survey of the Application of Modern Management Practices in the Cooperative Context* (Geneva: International Labour Organization, 1999), 47.

21. Kanigel, *One Best Way*, 482.

22. Kanigel, *One Best Way*, 11.

23. Nikolai Lenin, *The Soviets at Work* (New York: Rand School of Social Science, 1919). Kanigel, *One Best Way*, 524

24. Kanigel, *One Best Way*, 8.

25. M. Freeman, "Scientific Management: 100 Years Old; Poised for the Next Century," *SAM Advanced Management Journal* 61, no. 2 (1996): 35.

26. Richard J. Murnane and Stephen Hoffman, "Graduations on the Rise," EducationNext, http://educationnext.org/graduations-on-the-rise/; and "Education," PBS.com, http://www.pbs.org/fmc/book/3education1.htm.

27. Charles W. Eliot, *Educational Reform: Essays and Addresses* (New York: Century Co., 1901).

28. For an overview of the general debate, and the views of the Taylorists in particular, see Raymond E. Callahan, *Education and the Cult of Efficiency* (Chicago: University of Chicago Press, 1964).

29. Frederick T. Gates, "The Country School of To-Morrow," *Occasional Papers* 1 (1913): 6–10.

30. John Taylor Gatto, *The Underground History of American Education* (Odysseus Group, 2001), 222.

31. H. L. Mencken, "The Little Red Schoolhouse," *American Mercury*, April 1924, 504.

32. For biographical information on Thorndike, see Geraldine M. Joncich, *The Sane Positivist: A Biography of Edward L. Thorndike* (Middletown: Wesleyan University Press, 1968).

33. S. Tomlinson, "Edward Lee Thorndike and John Dewey on the Science of Education," *Oxford Review of Education* 23, no. 3 (1997): 365–383.

34. Callahan, *Education and the Cult of Efficiency*, 198.

35. Edward Thorndike, *Educational Psychology: Mental Work and Fatigue and Individual Differences and Their Causes* (New York: Columbia University, 1921), 236. Note: Like Galton, Thorndike was obsessed with ranking people. In his final book, *Human Nature and the Social Order* (1940), Thorndike proposed a system of moral scoring that could help society distinguish between superior and inferior citizens. An average man received a score of 100, while "Newton, Pasteur, Darwin, Dante, Milton, Bach, Beethoven, Leonardo da Vinci, and Rembrandt will count as 2000, and a vegetative idiot as about 1." In Thorndike's system of moral ranking, domesticated animals were assigned scores higher than human idiots.

36. Joncich, *The Sane Positivist*, 21–22.

37. Edward Thorndike, *Individuality* (Boston: Houghton Mifflin, 1911). Also see his approach to testing: Edward Thorndike, *An Introduction to the Theory of Mental and Social Measurements* (New York: Science Press, 1913).

38. Callahan, *Education and the Cult of Efficiency*, chap. 5.

39. Callahan, *Education and the Cult of Efficiency*, chap. 5.

40. Robert J. Marzano, "The Two Purposes of Teacher Evaluation," *Educational Leadership* 70, no. 3 (2012): 14–19, http://www.ascd.org/publications/educational-leadership/nov12/vol70/num03/The-Two-Purposes-of-Teacher-Evaluation.aspx; "Education Rankings," *U.S. News and World Report*, http://www.usnews.cm/rankings; "PISA 2012 Results," OECD, http://www.oecd.org/pisa/keyfindings/pisa–2012-results.htm.

41. Robert J. Murnane and Stephen Hoffman, "Graduations on the Rise," http://educationnext.org/graduations-on-the-rise/; "2015 Building a Grad Nation Report," Grad Nation, http://gradnation.org/report/2015-building-grad-nation-report.

42. Seth Godin, *We Are All Weird* (The Domino Project, 2011).

CHAPTER 3: OVERTHROWING THE AVERAGE

1. Peter Molenaar, interviewed by Todd Rose, August 18, 2014.

2. Molenaar, interview, 2014.

3. Frederic M. Lord and Melvin R. Novick, *Statistical Theories of Mental Test Scores* (Reading, MA: Addison-Wesley Publishing Co., 1968).

4. J. B. Kline, "Classical Test Theory: Assumptions, Equations, Limitations, and Item Analyses," in *Psychological Testing* (Calgary: University of Calgary, 2005), 91–106.

5. Lord and Novick, *Statistical Theories,* 27–28.

6. Lord and Novick, *Statistical Theories,* 29–32.

7. Lord and Novick, *Statistical Theories,* 32–35.

8. For a history and overview of ergodic theory, see Andre R. Cunha, "Understanding the Ergodic Hypothesis Via Analogies," *Physicae* 10, no. 10 (2013): 9–12; J. L. Lebowitz and O. Penrose, "Modern Ergodic Theory," *Physics Today* (1973): 23; Massimiliano Badino, "The Foundational Role of Ergodic Theory," *Foundations of Science* 11 (2006): 323–347; A. Patrascioiu, "The Ergodic Hypothesis: A Complicated Problem in Mathematics and Physics," *Los Alamos Science Special Issue* (1987): 263–279.

9. Ergodic theory was proved by the mathematician Birkhoff in 1931: G. D. Birkhoff, "Proof of the Ergodic Theorem," *Proceedings of the National Academy of Sciences of the United States of America* 17, no. 12 (1931): 656–660.

10. Peter C. M. Molenaar, "On the Implications of the Classical Ergodic Theorems: Analysis of Developmental Processes Has to Focus on Intra-Individual Variation," *Developmental Psychobiology* 50, no. 1 (2007): 60–69. Note: These two conditions are necessary and sufficient for Gaussian processes, which is what we have been discussing up to this point in the book. But they are not sufficient for general processes. Proving that a dynamic system is ergodic is exceedingly difficult and successfully carried out for only a small set of dynamic systems.

11. For example, see Bodrova et al., "Nonergodic Dynamics of Force-Free Granular Gases," arXiv:1501.04173 (2015); Thomas Scheby Kuhlman, *The Non-Ergodic Nature of Internal Conversion* (Heidelberg: Springer Science & Business Media, 2013); and Sydney Chapman et al., *The Mathematical Theory of Non-Uniform Gases* (Cambridge: Cambridge University Press, 1970). Note that some ideal gasses are ergodic; see, for instance, K. L. Volkovysskii and Y. G. Sinai, "Ergodic properties of an ideal gas

with an infinite number of degrees of freedom," *Functional Analysis and Its Applications,* no. 5 (1971): 185–187. Also note that ergodic theory was shown empirically to hold for diffusion in "Ergodic Theorem Passes the Test," *Physics World,* October 20, 2011, http://physicsworld.com/cws /article/news/2011/oct/20/ergodic-theorem-passes-the-test.

12. Peter Molenaar, interview, 2014. Also see Peter Molenaar et al., "Consequences of the Ergodic Theorems for Classical Test Theory, Factor Analysis, and the Analysis of Developmental Processes," in *Handbook of Cognitive Aging* (Los Angeles: SAGE Publications, 2008), 90–104.

13. A. Quetelet, *Lettres à S. A. R. Le Duc Régnant de Saxe Cobourg et Gotha, sur la théorie des probabilités, appliquée aux sciences morales et politique* (Brussels: Hayez, 1846), 136.

14. Peter Molenaar, "A Manifesto on Psychology as Idiographic Science: Bringing the Person Back into Scientific Psychology, This Time Forever," *Measurement* 2, no. 4 (2004): 201–218.

15. Molenaar, interview, 2014.

16. Molenaar, interview, 2014.

17. Molenaar, interview, 2014.

18. Molenaar, interview, 2014.

19. Rose et al., "Science of the Individual," 152–158.

20. Paul Van Geert, "The Contribution of Complex Dynamic Systems to Development," *Child Development Perspectives* 5, no. 4 (2011): 273–278.

21. Rose et al., "Science of the Individual," 152–158.

22. Anatole S. Dekaban, *Neurology of Infancy* (Baltimore: Williams & Wilkins, 1959), 63.

23. M. R. Fiorentino, *A Basis for Sensorimotor Development—Normal and Abnormal: The Influence of Primitive, Postural Reflexes on the Development and Distribution of Tone* (Springfield: Charles C. Thomas, 1981), 55; R. S. Illingworth, *The Development of the Infant and Young Child: Normal and Abnormal,* 3rd ed. (London: E. & S. Livingstone, 1966), 88; M. B. McGraw, "Neuromuscular Development of the Human Infant As Exemplified in the Achievement of Erect Locomotion," *Journal of Pediatrics* 17 (1940): 747–777; J. H. Menkes, *Textbook of Child Neurology* (Philadelphia: Lea & Febiger, 1980), 249; G. E. Molnar, "Analysis of Motor Disorder in Retarded Infants and Young Children," *American Journal of Mental Deficiency* 83 (1978): 213–222; A. Peiper, *Cerebral Function in Infancy and Childhood* (New York: Consultants Bureau, 1963), 213–215.

24. For a tribute to her work, see Karen E. Adolph and Beatrix Vereijken, "Esther Thelen (1941–2004)," *American Psychologist* 60, no. 9 (2005): 1032.

25. E. Thelen and D. M. Fisher, "Newborn Stepping: An Explanation for a 'Disappearing' Reflex," *Developmental Psychology* 18, no. 5 (1982): 760–775.

26. E. Thelen et al., "The Relationship Between Physical Growth and a Newborn Reflex," *Infant Behavior and Development* 7, no. 4 (1984): 479–493.

27. http://www.f22fighter.com/cockpit.htm.

CHAPTER 4: TALENT IS ALWAYS JAGGED

1. Robert Levering and Milton Moskowitz, "2007 100 Best Companies to Work for," Great Place to Work, http://www.greatplacetowork.net/best -companies/north-america/united-states/fortunes–100-best-companies -to-work-forr/439–2007.

2. Virginia A. Scott, *Google* (Westport: Greenwood Publishing Group, 2008), 61.

3. Steve Lohr, "Big Data, Trying to Build Better Workers," *New York Times*, April 20, 2013, http://www.nytimes.com/2013/04/21/technology/big -data-trying-to-build-better-workers.html?src=me&pagewanted=all& _r=1. See also Eric Schmidt and Jonathan Rosenberg, *How Google Works* (New York: Grand Central Publishing, 2014).

4. George Anders, *The Rare Find: How Great Talent Stands Out* (New York: Penguin, 2011), 3.

5. Leslie Kwoh, "'Rank and Yank' Retains Vocal Fans," *Wall Street Journal*, January 21, 2012, http://www.wsj.com/articles/SB100014240529702033 63504577186970064375222.

6. Ashley Goodall, interviewed by Todd Rose, April 17, 2015. See also, Marcus Buckingham and Ashley Goodall, "Reinventing Performance Management," *Harvard Business Review*, April 2015, https://hbr.org/2015/04 /reinventing-performance-management. Note: Goodall is now Senior Vice President for Leadership and Team Intelligence at Cisco Systems.

7. Kwoh, "'Rank and Yank.'"

8. For an overview of forced rankings, see Richard C. Grote, *Forced Ranking: Making Performance Management Work* (Cambridge: Harvard Business Press, 2005).

9. David Auerbach, "Tales of an Ex-Microsoft Manager: Outgoing CEO Steve Ballmer's Beloved Employee-Ranking System Made Me Secretive, Cynical and Paranoid," *Slate,* August 26, 2013, http://www.slate.com /articles/business/moneybox/2013/08/microsoft_ceo_steve_ballmer _retires_a_firsthand_account_of_the_company_s.html.

10. Kwoh, " 'Rank and Yank' " and Julie Bort, "This Is Why Some Microsoft Employees Still Fear the Controversial 'Stack Ranking' Employee Review System," *Business Insider,* August 27, 2014, http://www.businessinsider .com/microsofts-old-employee-review-system–2014–8.

11. Anders, *Rare Find,* 3–4. Also see Thomas L. Friedman, "How to Get a Job at Google," *New York Times,* February 22, 2014, http://www.nytimes .com/2014/02/23/opinion/sunday/friedman-how-to-get-a-job-at-google .html?_r=0.

12. Todd Carlisle, interviewed by Todd Rose, April 21, 2015.

13. Buckingham and Goodall, "Reinventing Performance Management."

14. Ashley Goodall, interviewed by Todd Rose, April 17, 2015.

15. Kurt Eichenwald, "Microsoft's Lost Decade," *Vanity Fair,* August 2012, http://www.vanityfair.com/news/business/2012/08/microsoft-lost-mojo -steve-ballmer.

16. Marcus Buckingham, "Trouble with the Curve? Why Microsoft Is Ditching Stack Rankings," *Harvard Business Review,* November 19, 2013, https://hbr.org/2013/11/dont-rate-your-employees-on-a-curve/.

17. Francis Galton, *Essays in Eugenics* (London: The Eugenics Education Society, 1909), 66.

18. For a broader discussion of one-dimensional thinking, see Paul Churchill, *A Neurocomputational Perspective: The Nature of Mind and the Structure of Science* (Cambridge, MA: MIT Press, 1989), 285–286; and Herbert Marcuse, *One-Dimensional Man: Studies in the Ideology of Advanced Industrial Society,* 2nd ed. (London: Routledge, 1991).

19. Daniels, *The "Average Man"?,* 3.

20. William F. Moroney and Margaret J. Smith, *Empirical Reduction in Potential User Population as the Result of Imposed Multivariate Anthropometric Limits* (Pensacola, FL: Naval Aerospace Medical Research Laboratory, 1972), NAMRL–1164.

21. David Berri and Martin Schmidt, *Stumbling on Wins* (Bonus Content Edition) (New York: Pearson Education, 2010), Kindle Edition, chap. 2.

22. David Berri, "The Sacrifice LeBron James' Teammates Make to Play Alongside Him," *Time,* October 16, 2014, http://time.com/3513970/lebron

-james-shot-attempts-scoring-totals/; also see Henry Abbott, "The Robots Are Coming, and They're Cranky," *ESPN,* March 17, 2010, http://espn .go.com/blog/truehoop/post/_/id/14349/the-robots-are-coming-and -theyre-cranky.

23. David Berri, "Bad Decision Making Is a Pattern with the New York Knicks," *Huffington Post,* May 14, 2015, http://www.huffingtonpost.com /david-berri/bad-decision-making-is-a-_b_7283466.html.

24. Berri and Schmidt, *Stumbling on Wins,* chap. 2; also see David Berri, "The Sacrifice LeBron James' Teammates Make to Play Alongside Him," *Time.com,* October 16, 2014, http://time.com/3513970/lebron-james -shot-attempts-scoring-totals/.

25. David Friedman, "Pro Basketball's 'Five-Tool' Players," *20 Second Time-out,* March 25, 2009, http://20secondtimeout.blogspot.com/2009/03 /pro-basketballs-five-tool-players_25.html.

26. Dean Oliver, *Basketball on paper: rules and tools for performance analysis* (Potomac Books, 2004), 63–64. For qualitative insights about building successful teams, see Mike Krzyzewski, *The Gold Standard: Building a World-Class Team* (New York, Business Plus, 2009).

27. Berri, "Bad Decision Making."

28. D. Denis, "The Origins of Correlation and Regression: Francis Galton or Auguste Bravais and the Error Theorists," *History and Philosophy of Psychology Bulletin* 13 (2001): 36–44.

29. Francis Galton, "Co-relations and Their Measurement, Chiefly from Anthropometric Data," *Proceedings of the Royal Society of London* 45, no. 273–279 (1888): 135–145.

30. Technically correlations range from −1.00 to +1.00 with the sign indicating the direction of the relationship. Since the point I am trying to make here is about the strength of the relationship, I chose to present it as 0 to 1 for the sake of clarity.

31. "Five Questions About the Dow That You Always Wanted to Ask," Dow Jones Indexes, February 2012, https://www.djindexes.com/mdsidx /downloads/brochure_info/Five_Questions_Brochure.pdf.

32. William F. Moroney and Margaret J. Smith, *Empirical Reduction in Potential User Population as the Result of Imposed Multivariate Anthropometric Limits* (Pensacola, FL: U.S. Department of the Navy, 1972), NAMRL-1164. The data analyzed in the study is from E. C. Gifford, *Compilation of Anthropometric Measures on US Naval Pilot* (Philadelphia: U.S. Department of the Navy, 1960), NAMC-ACEL–437. For practi-

cal consequences of the lack of fit, see George T. Lodge, *Pilot Stature in Relation to Cockpit Size: A Hidden Factor in Navy Jet Aircraft Accidents* (Norfolk, VA: Naval Safety Center, 1964).

33. Francis Galton, "Mental Tests and Measurements," *Mind* 15, no. 59 (1890): 373–381.

34. For biographical information, see W. B. Pillsbury, *Biographical Memoir of James McKeen Cattell 1860–1944* (Washington, DC: National Academy of the Sciences, 1947); and M. M. Sokal, "Science and James McKeen Cattell, 1894–1945," *Science* 209, no. 4452 (1980): 43–52.

35. James McKeen Cattell and Francis Galton, "Mental Tests and Measurements," *Mind* 13 (1890): 37–51; and James McKeen Cattell and Livingstone Farrand, "Physical and Mental Measurements of the Students of Columbia University," *Psychological Review* 3, no. 6 (1896): 618. Also see Michael M. Sokal, "James McKeen Cattell and Mental Anthropometry: Nineteenth-Century Science and Reform and the Origins of Psychological Testing," in *Psychological Testing and American Society, 1890–1930*, ed. Michael Sokal (New Brunswick: Rutgers University Press, 1987).

36. The results were analyzed and published as part of the doctoral dissertation of Cattell's student, Clark Wissler. See Clark Wissler, "The Correlation of Mental and Physical Tests," *Psychological Review: Monograph Supplements* 3, no. 6 (1901): i.

37. Wissler, "Correlation of Mental and Physical Tests," i.

38. Charles Spearman, " 'General Intelligence,' Objectively Determined and Measured," *American Journal of Psychology* 15, no. 2 (1904): 201–292.

39. For a terrific study that shows not only the fact of jaggedness in individuals, but also that individuals differ in the amount of their jaggedness, see C. L. Hull, "Variability in Amount of Different Traits Possessed by the Individual," *Journal of Educational Psychology* 18, no. 2 (February 1, 1927): 97–106. For a more current study, see Laurence M. Binder et al., "To Err Is Human: 'Abnormal' Neuropsychological Scores and Variability Are Common in Healthy Adults," *Archives of Clinical Neuropsychology* 24, no. 1 (2009): 31–46.

40. G. C. Cleeton, and Frederick B. Knight, "Validity of Character Judgments Based on External Criteria," *Journal of Applied Psychology* 8, no. 2 (1924): 215.

41. For a discussion of his father's study, see Robert L. Thorndike and Elizabeth Hagen, *Ten Thousand Careers* (New York: John Wiley & Sons, 1959). Note: To any reader familiar with his views it will seem strange to attribute to Thorndike a one-dimensional view of intelligence, since

he was consistently arguing intelligence was multidimensional (abstract, social, and mechanical) and was one of Spearman's biggest critics. However, he did believe there was an innate component that applied to your ability to learn and that it had to do with your neural ability to form connections.

42. David Wechsler, *Wechsler Adult Intelligence Scale—Fourth Edition (WAIS–IV)* (San Antonio, TX: NCS Pearson, 2008).

43. Wayne Silverman et al., "Stanford-Binet and WAIS IQ Differences and Their Implications for Adults with Intellectual Disability (aka Mental Retardation)," *Intelligence* 38, no. 2 (2010): 242–248.

44. This extends to all traits that we typically measure. See Hull, "Variability in Amount of Different Traits," 97–106.

45. Jerome M. Sattler and Joseph J. Ryan, *Assessment with the WAIS-IV* (La Mesa, CA: Jerome M. Sattler Publisher, 2009). For more on the inherently jagged nature of intelligence, see Adam Hampshire et al., "Fractionating Human Intelligence," *Neuron*, December 10 (2012): 1–13.

46. Sergio Della Sala et al., "Pattern Span: A Tool for Unwelding Visuo-Spatial Memory," *Neuropsychologia* 37, no. 10 (1999): 1189–1199.

47. Jennifer L. Kobrin et al., *Validity of the SAT for Predicting First-Year College Grade Point Average* (New York: College Board, 2008).

48. Steve Jost, "Linear Correlation," course document, IT 223, DePaul University, 2010, http://condor.depaul.edu/sjost/it223/documents/correlation.htm.

49. Todd Carlisle, interviewed by Todd Rose, April 21, 2015.

50. Carlisle, interview, 2015.

51. Todd Carlisle, interview, 2015; also see Saul Hansell, "Google Answer to Filling Jobs Is an Algorithm," *New York Times,* January 3, 2007, http://www.nytimes.com/2007/01/03/technology/03google.html?pagewanted=1&_r=2&; for similar insights about Todd Carlisle's thinking, approach, and results, see Anders, *Rare Find.*

52. Carlisle, interview, 2015.

53. Carlisle, interview, 2015. See also Saul Hansell, "Google Answer to Filling Jobs Is an Algorithm," *New York Times,* January 3, 2007, http://www.nytimes.com/2007/01/03/technology/03google.html?pagewanted=2&_r=0.

54. Employee numbers were taken from "Google," *Wikipedia,* June 19, 2015, http://en.wikipedia.org/wiki/Google; and "IGN," *Wikipedia,* June 13, 2015, http://en.wikipedia.org/wiki/IGN. Yearly sales num-

bers were taken from: "Google," *Forbes,* http://www.forbes.com
/companies/google/; and "j2 Global," *Forbes,* http://www.forbes.com
/companies/j2-global/), with IGN numbers based on the parent company, j2 Global.

55. E. B. Boyd, "Silicon Valley's New Hiring Strategy," *Fast Company,* October 20, 2011, http://www.fastcompany.com/1784737/silicon-valleys-new
-hiring-strategy.

56. http://www.ign.com/code-foo/2015/.

57. Boyd, "Silicon Valley."

58. Boyd, "Silicon Valley."

59. "GRE," ETS, http://www.ets.org/gre.

CHAPTER 5: TRAITS ARE A MYTH

1. Francis Galton, "Measurement of Character," reprinted in *Fortnightly
Review* 42 (1884): 180.

2. L. Rowell Huesmann and Laramie D. Taylor, "The Role of Media Violence in Violent Behavior," *Annual Review of Public Health* 27 (2006):
393–415. For an overview of the situationist perspective see Lee Ross and
Richard E. Nisbett, *The Person and the Situation: Perspectives of Social
Psychology* (London: Pinter & Martin Publishers, 2011).

3. Quetelet, *Sur l'homme* (1942) 108 (English edition).

4. Stanley Milgram, "Behavioral Study of Obedience," *Journal of Abnormal
and Social Psychology* 67, no. 4 (1963): 371.

5. Milgram, "Behavioral Study of Obedience."

6. Douglas T. Kenrick and David C. Funder, "Profiting from Controversy:
Lessons from the Person-Situation Debate," *American Psychologist* 43, no.
1 (1988): 23.

7. "Understanding the Personality Test Industry," Psychometric Success,
http://www.psychometric-success.com/personality-tests/personality
-tests-understanding-industry.htm; Lauren Weber, "Today's Personality
Tests Raise the Bar for Job Seekers," *Wall Street Journal,* April 14, 2015,
http://www.wsj.com/articles/a-personality-test-could-stand-in-the-way
-of-your-next-job–1429065001.

8. Drake Baer, "Why the Myers-Briggs Personality Test Is Misleading,
Inaccurate, and Unscientific," *Business Insider,* June 18, 2014, http://

www.businessinsider.com/myers-briggs-personality-test-is
-misleading–2014–6; and Lillian Cunningham, "Myers-Briggs: Does It
Pay to Know Your Type?" *Washington Post,* December 14, 2012, http://
www.washingtonpost.com/national/on-leadership/myers-briggs
-does-it-pay-to-know-your-type/2012/12/14/eaed51ae–3fcc–11e2-bca3
-aadc9b7e29c5_story.html.

9. Salesforce.com, "How to Use the Enneagram in Hiring Without Using a
Candidate's Enneatype," The *Enneagram in Business,* October 25, 2012,
http://theenneagraminbusiness.com/organizations/salesforce-com-how
-to-use-the-enneagram-in-hiring-without-using-a-candidates-enneatype/.

10. Lawrence W. Barsalou et al., "On the Vices of Nominalization and the
Virtues of Contextualizing," in *The Mind in Context,* ed. Batja Mesquita
et al. (New York: Guilford Press, 2010), 334–360; Susan A. Gelman,
The Essential Child: Origins of Essentialism in Everyday Thought (Oxford:
Oxford University Press, 2003); David L. Hull, "The Effect of Essential-
ism on Taxonomy—Two Thousand Years of Stasis (I)," *British Journal
for the Philosophy of Science* (1965): 314–326; and Douglas L. Medin and
Andrew Ortony, "Psychological Essentialism," *Similarity and Analogical
Reasoning* 179 (1989): 195.

11. John Tierney, "Hitting It Off, Thanks to Algorithms of Love," *New
York Times,* January 29, 2008, http://www.nytimes.com/2008/01/29
/science/29tier.html?_r=0; and "28 Dimensions of Compatibility," http://
www.eharmony.com/why/dating-relationship-compatibility/.

12. J. McV. Hunt, "Traditional Personality Theory in Light of Recent Evidence," *American Scientist* 53, no. 1 (1965): 80–96. Walter Mischel,
"Continuity and Change in Personality," *American Psychologist* 24, no. 11
(1969): 1012; and Walter Mischel, *Personality and Assessment* (New York:
Psychology Press, 2013).

13. Erik E. Noftle and Richard W. Robins, "Personality Predictors of Academic Outcomes: Big Five Correlates of GPA and SAT Scores," *Journal of
Personality and Social Psychology* 93, no. 1 (2007): 116; and Ashley S. Hol-
land and Glenn I. Roisman, "Big Five Personality Traits and Relationship Quality: Self-Reported, Observational, and Physiological Evidence,"
Journal of Social and Personal Relationships 25, no. 5 (2008): 811–829.

14. "Yuichi Shoda, Ph.D.," University of Washington Psychology Department Directory, http://web.psych.washington.edu/directory/areapeople
.php?person_id=85.

15. Yuichi Shoda, interviewed by Todd Rose, November 19, 2014.

16. Shoda, interview, 2014.

17. "Research," Wediko Children's Services, http://www.wediko.org/research
.html.

18. Yuichi Shoda et al., "Intraindividual Stability in the Organization and
Patterning of Behavior: Incorporating Psychological Situations into the
Idiographic Analysis of Personality," *Journal of Personality and Social Psychology* 67, no. 4 (1994): 674.

19. Shoda et al., "Intraindividual Stability in the Organization and Patterning of Behavior."

20. Shoda et al., "Intraindividual Stability in the Organization and Patterning of Behavior."

21. Lisa Feldman Barrett et al., "The Context Principle," in *The Mind in Context,* ed. Batja Mesquita, Lisa Feldman Barrett, and Eliot R. Smith (New York: Guildford Press, 2010), chap. 1; Walter Mischel, "Toward an Integrative Science of the Person," *Annual Review of Psychology* 55 (2004): 1–22; Yuichi Shoda, Daniel Cervone, and Geraldine Downey, eds., *Persons in Context: Building a Science of the Individual* (New York: Guilford Press, 2007); and Robert J. Sternberg and Richard K. Wagner, *Mind in Context: Interactionist Perspectives on Human Intelligence* (Cambridge: Cambridge University Press, 1994).

22. Shoda et al., *Persons in Context.*

23. Lara K. Kammrath et al., "Incorporating If . . . Then . . . Personality Signatures in Person Perception: Beyond the Person-Situation Dichotomy," *Journal of Personality and Social Psychology* 88, no. 4 (2005): 605; Batja Mesquita, Lisa Feldman Barrett, and Eliot R. Smith, eds., *The Mind in Context* (New York: Guilford Press, 2010); Sternberg and Wagner, *Mind in Context;* and Donna D. Whitsett and Yuichi Shoda, "An Approach to Test for Individual Differences in the Effects of Situations Without Using Moderator Variables," *Journal of Experimental Social Psychology* 50, no. C (January 1, 2014): 94–104.

24. For biographical information see Raymond P. Morris, "Hugh Hartshorne, "1885–1967," *Religious Education* 62, no. 3 (1968): 162.

25. Marvin W. Berkowitz and Melinda C. Bier, "Research-Based Character Education," *Annals of the American Academy of Political and Social Science* 591, no. 1 (2004): 72–85.

26. Hartshorne and May, *Studies, Vol. 1: Studies in Deceit,* 47–103.

27. Hartshorne and May, *Studies, Vol. 1: Studies in Deceit.* Also see John M. Doris, *Lack of Character: Personality and Moral Behavior* (Cambridge: Cambridge University Press, 2002).

28. Hartshorne, May, and Shuttleworth, *Studies, Vol. III: Studies in the Orga-*

nization of Character (1930): 291. Note: In the original study one of the students was a boy and the other was a girl, but for the purposes of illustration I have chosen to talk about each one as a girl so that the focus would be on the character profiles rather than gender.

29. Hartshorne, May, and Shuttleworth, *Studies, Vol. III: Studies in the Organization of Character,* 287.

30. For a recent example, see Mark Prigg, "Self Control Is the Most Important Skill a Parent Can Teach Their Child, Says Study," *Daily Mail,* April 14, 2015, http://www.dailymail.co.uk/sciencetech/article–3038807 /Self-control-important-thing-parent-teach-children-Study-says-major -influence-child-s-life.html.

31. For an overview of the subject, see the recent book from the originator of the task, Walter Mischel, *The Marshmallow Test* (New York: Random House, 2014). For details of the task, see "Delaying Gratification," in "What You Need to Know about Willpower: The Psychological Science of Self-Control," American Psychological Association, https://www .apa.org/helpcenter/willpower-gratification.pdf; and "Stanford Marshmallow Experiment," *Wikipedia,* June 13, 2015, https://en.wikipedia.org /wiki/Stanford_marshmallow_experiment.

32. Walter Mischel et al., "The Nature of Adolescent Competencies Predicted by Preschool Delay of Gratification," *Journal of Personality and Social Psychology* 54, no. 4 (1988): 687; Walter Mischel et al., "Cognitive and Attentional Mechanisms in Delay of Gratification," *Journal of Personality and Social Psychology* 21, no. 2 (1972): 204.

33. Yuichi Shoda et al., "Predicting Adolescent Cognitive and Self-Regulatory Competencies from Preschool Delay of Gratification: Identifying Diagnostic Conditions," *Developmental Psychology* 26, no. 6 (1990): 978. See also Walter Mischel and Nancy Baker, "Cognitive Appraisals and Transformations in Delay Behavior," *Journal of Personality and Social Psychology* 31, no. 2 (1975): 254; Walter Mischel et al., "Delay of Gratification in Children," *Science* 244, no. 4907 (1989): 933–938; Walter Mischel et al., "'Willpower' over the Life Span: Decomposing Self-Regulation," *Social Cognitive and Affective Neuroscience* (2010); Tanya R. Schlam et al., "Preschoolers' Delay of Gratification Predicts Their Body Mass 30 Years Later," *Journal of Pediatrics* 162, no. 1 (2013): 90–93; and Inge-Marie Eigsti, "Predicting Cognitive Control from Preschool to Late Adolescence and Young Adulthood," *Psychological Science* 17, no. 6 (2006): 478–484.

34. B. J. Casey et al., "Behavioral and Neural Correlates of Delay of Gratification 40 Years Later," *Proceedings of the National Academy of Sciences* 108, no. 36 (2011): 14998–15003.

35. Louise Eckman, "Behavior Problems: Teaching Young Children Self-Control Skills," National Mental Health and Education Center, http://www.nasponline.org/resources/handouts/behavior%20template.pdf.

36. Martin Henley, *Teaching Self-Control: A Curriculum for Responsible Behavior* (Bloomington, IN: National Educational Service, 2003); and "Self Control," Character First Education, http://characterfirsteducation.com/c/curriculum-detail/2039081.

37. For a discussion, see Jacoba Urist, "What the Marshmallow Test Really Teaches About Self-Control," *Atlantic,* September 24, 2014, http://www.theatlantic.com/health/archive/2014/09/what-the-marshmallow-test-really-teaches-about-self-control/380673/.

38. Shoda, interview, 2014.

39. For more information about Dr. Kidd's work, see "Celeste Kidd," University of Rochester, Brain & Cognitive Sciences, http://www.bcs.rochester.edu/people/ckidd/.

40. Celeste Kidd, interviewed by Todd Rose, June 12, 2015; see also "The Marshmallow Study Revisited," University of Rochester, October 11, 2012, http://www.rochester.edu/news/show.php?id=4622.

41. Kidd et al., "Rational Snacking: Young Children's Decision-Making on the Marshmallow Task Is Moderated by Beliefs About Environmental Reliability," *Cognition* 126, no. 1 (2013): 109–114.

42. Kidd et al., "Rational Snacking."

43. "What We Do," Adler Group, http://louadlergroup.com/about-us/what-we-do/.

44. Lou Adler, interviewed by Todd Rose, March 27, 2015.

45. Adler, interview, 2015; for an overview of Performance-Based Hiring, see Lou Adler, *Hire with Your Head: Using Performance-Based Hiring to Build Great Teams* (Hoboken: John Wiley & Sons, 2012).

46. Adler, interview, 2015.

47. Adler, interview, 2015.

48. Dr. Matthew Partridge, "Callum Negus-Fancey: 'Put People and Talent First,'" *MoneyWeek,* January 22, 2015, http://moneyweek.com/profile-of-entrepreneur-callum-negus-fancey/.

49. Callum Negus-Fancey, interviewed by Todd Rose, April 3, 2015.

50. Negus-Fancey, interview, 2015.

51. Negus-Fancey, interview, 2015.

52. Adler, interview, 2015.

CHAPTER 6: WE ALL WALK THE ROAD LESS TRAVELED

1. Arnold Gesell, "Developmental Schedules," in *The Mental Growth of the Pre-School Child: A Psychological Outline of Normal Development from Birth to the Sixth Year, Including a System of Developmental Diagnosis* (New York, NY: Macmillan, 1925).

2. Robert Kanigel, *The One Best Way: Frederick Winslow Taylor and the Enigma of Efficiency* (Cambridge: MIT Press Books, 2005).

3. Raymond E. Callahan, *Education and the Cult of Efficiency* (Chicago: University of Chicago Press, 1964).

4. E. Thelen and K. E. Adolph, "Arnold L. Gesell: The Paradox of Nature and Nurture," *Developmental Psychology* 28, no. 3 (1992): 368–380; Laura Sices, "Use of Developmental Milestones in Pediatric Residency Training and Practice: Time to Rethink the Meaning of the Mean," *Journal of Developmental and Behavioral Pediatrics* 28, no. 1 (2007): 47; K. E. Adolph and S. R. Robinson, "The Road to Walking: What Learning to Walk Tells Us About Development," in *Oxford Handbook of Developmental Psychology*, ed. P. Zelazo (New York: Oxford University Press, 2013); and "Child Growth Standards: Motor Development Milestones," *World Health Organization*, http://www.who.int/childgrowth/standards/motor_milestones/en/.

5. For information about Dr. Karen Adolph and her work, see her lab website: http://psych.nyu.edu/adolph/.

6. Karen E. Adolph et al., "Learning to Crawl," *Child Development* 69, no. 5 (1998): 1299–1312.

7. Adolph et al., "Learning to Crawl."

8. Adolph et al., "Learning to Crawl."

9. Karen Adolph, interviewed by Todd Rose, June 13, 2015.

10. "Discovery: Will Baby Crawl?" *National Science Foundation*, July 21, 2004, https://www.nsf.gov/discoveries/disc_summ.jsp?cntn_id=103153.

11. Kate Gammon, "Crawling: A New Evolutionary Trick?" *Popular Science*, November 1, 2013, http://www.popsci.com/blog-network/kinderlab/crawling-new-evolutionary-trick.

12. "David Tracer, Ph.D." University of Colorado Denver Fulbright Scholar Recipients, http://www.ucdenver.edu/academics/InternationalPrograms/oia/fulbright/recipients/davidtracer/Pages/default.aspx; Kate Wong, "Hitching a Ride," *Scientific American* 301, no. 1 (2009): 20–23; "Discovery: Will Baby Crawl?"

13. "What Are the Key Statistics About Colorectal Cancer?" American Cancer Society, http://www.cancer.org/cancer/colonandrectumcancer /detailedguide/colorectal-cancer-key-statistics.

14. Eric R. Fearon and Bert Vogelstein, "A Genetic Model for Colorectal Tumorigenesis," *Cell* 61, no. 5 (1990): 759–767.

15. Gillian Smith et al., "Mutations in APC, Kirsten-ras, and p53— Alternative Genetic Pathways to Colorectal Cancer," *Proceedings of the National Academy of Sciences* 99, no. 14 (2002): 9433–9438; Massimo Pancione et al., "Genetic and Epigenetic Events Generate Multiple Pathways in Colorectal Cancer Progression," *Pathology Research International* 2012 (2012); Sylviane Olschwang et al., "Alternative Genetic Pathways in Colorectal Carcinogenesis," *Proceedings of the National Academy of Sciences* 94, no. 22 (1997): 12122–12127; and Yu-Wei Cheng et al., "CpG Island Methylator Phenotype Associates with Low-Degree Chromosomal Abnormalities in Colorectal Cancer," *Clinical Cancer Research* 14, no. 19 (2008): 6005–6013.

16. Daniel L. Worthley and Barbara A. Leggett, "Colorectal Cancer: Molecular Features and Clinical Opportunities," *Clinical Biochemist Reviews* 31, no. 2 (2010): 31.

17. Kenneth I. Howard et al., "The Dose–Effect Relationship in Psychotherapy," *American Psychologist* 41, no. 2 (1986): 159; Wolfgang Lutz et al., "Outcomes Management, Expected Treatment Response, and Severity-Adjusted Provider Profiling in Outpatient Psychotherapy," *Journal of Clinical Psychology* 58, no. 10 (2002): 1291–1304.

18. Jeffrey R. Vittengl et al., "Nomothetic and Idiographic Symptom Change Trajectories in Acute-Phase Cognitive Therapy for Recurrent Depression," *Journal of Consulting and Clinical Psychology* 81, no. 4 (2013): 615.

19. Three papers discussing the issue of equifinality: As it relates to development, see Dante Cicchetti and Fred A. Rogosch, "Equifinality and Multifinality in Developmental Psychopathology," *Development and Psychopathology* 8, no. 04 (1996): 597–600; leadership development, see Marguerite Schneider and Mark Somers, "Organizations as Complex Adaptive Systems: Implications of Complexity Theory for Leadership Research," *Leadership Quarterly* 17, no. 4 (2006): 351–365; and hydrology, see Keith Beven, "A Manifesto for the Equifinality Thesis," *Journal of Hydrology* 320, no. 1 (2006): 18–36.

20. Kurt W. Fischer and Thomas R. Bidell, "Dynamic Development of Action and Thought," in *Handbook of Child Psychology* (Hoboken, NJ: John Wiley & Sons, 2006); and Kathleen M. Eisenhardt and Jeffrey A.

Martin, "Dynamic Capabilities: What Are They?" *Strategic Management Journal* 21, no. 10–11 (2000): 1105–1121.

21. Edward L. Thorndike, "Memory for Paired Associates," *Psychological Review* 15, no. 2 (1908): 122.

22. Edward L. Thorndike, *The Human Nature Club: An Introduction to the Study of Mental Life* (New York: Longmans, Green, and Company, 1901), chap. 1.

23. Edward L. Thorndike, "Measurement in Education," *The Teachers College Record* 22, no. 5 (1921): 371–379; and Linda Mabry, "Writing to the Rubric: Lingering Effects of Traditional Standardized Testing on Direct Writing Assessment," *Phi Delta Kappan* 80, no. 9 (1999): 673.

24. Raiann Rahman, "The Almost Standardized Aptitude Test: Why Extra Time Shouldn't Be an Option on Standardized Testing," *Point of View*, October 18, 2013, http://www.bbnpov.com/?p=1250.

25. For biographical and background information on Benjamin Bloom and his career, see Thomas R. Guskey, *Benjamin S. Bloom: Portraits of an Educator* (Lanham, MD: R&L Education, 2012); and Elliot W. Eisner, "Benjamin Bloom," *Prospects* 30, no. 3 (2000): 387–395.

26. Benjamin S. Bloom, "Time and Learning," *American Psychologist* 29, no. 9 (1974): 682; and Benjamin S. Bloom, *Human Characteristics and School Learning* (New York: McGraw-Hill, 1976).

27. While Bloom rightly gets credit for the ideas, it is worth noting that the seminal studies were done by two of his doctoral students, Joanne Anania (Joanne Anania, "The Influence of Instructional Conditions on Student Learning and Achievement," *Evaluation in Education* 7, no. 1 [1983]: 1–92) and Arthur Burke (Arthur Joseph Burke, "Students' Potential for Learning Contrasted Under Tutorial and Group Approaches to Instruction" [Ph.D. diss., University of Chicago, 1983]).

28. In these studies there is an additional experimental condition examined—group-based mastery learning—that is not relevant to this particular discussion.

29. Benjamin S. Bloom, "The 2 Sigma Problem: The Search for Methods of Group Instruction as Effective as One-to-One Tutoring," *Educational Researcher* (1984): 4–16.

30. Chen-Lin C. Kulik et al., "Effectiveness of Mastery Learning Programs: A Meta-Analysis," *Review of Educational Research* 60, no. 2 (1990): 265–299.

31. Bloom, "2 Sigma Problem," 4–16.

32. Khan Academy, https://www.khanacademy.org/; and "Khan Academy," *Wikipedia*, June 3, 2015, https://en.wikipedia.org/wiki/Khan_Academy.

33. Anya Kamenetz, "A Q&A with Salman Khan, Founder of Khan Academy," *Fast Company*, November 21, 2013, http://live.fastcompany.com/Event/A_QA_With_Salman_Khan.

34. "A Personalized Learning Resource for All Ages," Khan Academy, https://www.khanacademy.org/about.

35. "Salman Khan," TED, https://www.ted.com/speakers/salman_khan.

36. "Khan," TED.

37. Arnold Gesell, "Arnold Gesell," *Psychiatric Research Reports* 13 (1960): 1–9.

38. Arnold Gesell and Catherine Strunk Amatruda, *The Embryology of Behavior: The Beginnings of the Human Mind* (New York: Harper & Brothers, 1945); Arnold Gesell, *The Ontogenesis of Infant Behavior* (New York: Wiley & Sons, 1954); Gesell, *Mental Growth of the Pre-School Child;* Arnold Gesell, *Infancy and Human Growth* (New York: MacMillan, 1928); Arnold Gesell and Helen Thompson, *Infant Behavior: Its Genesis and Growth* (New York: McGraw-Hill, 1934); Arnold Gesell, *How a Baby Grows* (New York: Harper & Brothers, 1945); Thomas C. Dalton, "Arnold Gesell and the Maturation Controversy," *Integrative Physiological & Behavioral Science* 40, no. 4 (2005): 182–204; and Fredric Weizmann and Ben Harris, "Arnold Gesell: The Maturationist," in *Portraits of Pioneers in Developmental Psychology* 7 (New York: Psychology Press, 2012).

39. Gesell, "Developmental Schedules;" and Gesell and Thompson, "Infant Behavior."

40. Gesell, "Developmental Schedules," as cited in Adolph et al., "Learning to Crawl." See also Adolph, Karen E., and Sarah E. Berger, "Motor Development," *Handbook of Child Psychology* (2006).

41. Gesell and Thompson, *Infant Behavior: Its Genesis and Growth,* chap. 3.

42. Weizmann and Harris, "Gesell: The Maturationist," 1.

43. Gesell and Amatruda, *Developmental Diagnosis* (New York: Harper, 1947).

44. Gesell and Amatruda, *Developmental Diagnosis,* 361.

45. Arnold Gesell, "Reducing the Risks of Child Adoption," *Child Welfare League of America Bulletin* 6, no. 3 (1927); and Ellen Herman, "Families Made by Science: Arnold Gesell and the Technologies of Modern Child Adoption," *Isis* (2001): 684–715.

46. Thelen and Adolph, "Gesell: Paradox of Nature and Nurture," 368–380.

47. Arlene Eisenberg et al., *What to Expect When You're Expecting* (New York: Simon & Schuster, 1996); and Heidi Murkoff et al., *What to Expect the First Year* (New York: Workman Publishing, 2009).

48. Thomas R. Bidell and Kurt W. Fischer, "Beyond the Stage Debate: Action, Structure, and Variability in Piagetian Theory and Research," *Intellectual Development* (1992): 100–140.

49. Rose et al., "The Science of the Individual," 152–158; L. Todd Rose and Kurt W. Fischer, "Dynamic Development: A Neo-Piagetian Approach," in *The Cambridge Companion to Piaget* (Cambridge: Cambridge University Press, 2009): 400; L. Todd Rose and Kurt W. Fischer, "Intelligence in Childhood," in *The Cambridge Handbook of Intelligence* (Cambridge: Cambridge University Press, 2011): 144–173.

50. "Kurt W. Fischer," *Wikipedia*, May 17, 2015, https://en.wikipedia.org/wiki /Kurt_W._Fischer.

51. For an overview of his work, see Kurt W. Fischer and Thomas R. Bidell, "Dynamic Development of Action and Thought," in *Handbook of Child Psychology*, 6th ed. (Hoboken, NJ: Wiley, 2006).

52. Catharine C. Knight and Kurt W. Fischer, "Learning to Read Words: Individual Differences in Developmental Sequences," *Journal of Applied Developmental Psychology* 13, no. 3 (1992): 377–404.

53. Kurt Fischer, interviewed by Todd Rose, August 14, 2014.

54. Knight and Fischer, "Learning to Read Words."

55. Knight and Fischer, "Learning to Read Words."

56. Fischer, interview, 2014.

57. Tania Rabesandratana, "Waltz to Excellence," *Science,* August 7, 2014, http://sciencecareers.sciencemag.org/career_magazine/previous_issues /articles/2014_08_07/caredit.a1400200.

58. Rabesandratana, "Waltz to Excellence."

59. Rabesandratana, "Waltz to Excellence."

60. Rabesandratana, "Waltz to Excellence."

61. "Characteristics of Remedial Students," Colorado Community College System, http://highered.colorado.gov/Publications/General/Strategic Planning/Meetings/Resources/Pipeline/Pipeline_100317_Remedial _Handout.pdf; and "Beyond the Rhetoric: Improving College Readi-

ness Through Coherent State Policy," http://www.highereducation.org/reports/college_readiness/gap.shtml.

62. CLEP (College Level Examination Program), https://clep.collegeboard.org/.

CHAPTER 7: WHEN BUSINESSES COMMIT TO INDIVIDUALITY

1. Victor Lipman, "Surprising, Disturbing Facts from the Mother of All Employment Engagement Surveys," *Forbes,* September 23, 2013, http://www.forbes.com/sites/victorlipman/2013/09/23/surprising-disturbing-facts-from-the-mother-of-all-employee-engagement-surveys/.

2. "Glassdoor's Employee's Choice Awards 2015: Best Places to Work 2015," *Glassdoor,* http://www.glassdoor.com/Best-Places-to-Work-LST_KQ0,19.htm; Rich Duprey, "6 Reasons Costco Wholesale Is the Best Retailer to Work For," *The Motley Fool,* December 13, 2014, http://www.fool.com/investing/general/2014/12/13/6-reasons-costco-wholesale-is-the-best-retailer-to.aspx; and "Top Companies for Compensation & Benefits 2014," *Glassdoor,* http://www.glassdoor.com/Top-Companies-for-Compensation-and-Benefits-LST_KQ0,43.htm.

3. Duprey, "6 Reasons."

4. Jim Sinegal, interviewed by Todd Rose, April 8, 2015.

5. Duprey, "6 Reasons"; "Jim Sinegal on Costco's 'Promote from Within' Strategy and Why It Needs to Think Like a Small Company," *The Motley Fool,* June 21, 2012, http://www.fool.com/investing/general/2012/06/21/jim-sinegal-on-costcos-promote-from-within-strateg.aspx.

6. Annette Alvarez-Peters, interviewed by Todd Rose (e-mail), May 5, 2015. Note: Alvarez-Peters started out at Price Club, which merged with Costco in 1993.

7. "Annette Alvarez-Peters," *Taste Washington,* http://tastewashington.org/annette-alvarez-peters/.

8. "The Decanter Power List 2013," *Decanter,* July 2, 2013, http://www.decanter.com/wine-pictures/the-decanter-power-list–2013–14237/.

9. Sinegal, interview, 2015.

10. Christ Horst, "An Open Letter to the President and CEO of Costco," *Smorgasblurb,* August 4, 2010, http://www.smorgasblurb.com/2010/08/an-open-letter-to-costco-executives/.

11. Sinegal, interview, 2015.

12. Adam Levine-Weinberg, "Why Costco Stock Keeps Rising," *The Motley Fool,* May 21, 2013, http://www.fool.com/investing/general/2013/05/21 /why-costco-stock-keeps-rising.aspx.

13. Andres Cardenal, "Costco vs. Wal-Mart: Higher Wages Mean Superior Returns for Investors," *The Motley Fool,* March 12, 2014, http://www .fool.com/investing/general/2014/03/12/costco-vs-wal-mart-higher -wages-mean-superior-retu.aspx.

14. Duprey, "6 Reasons;" and Jeff Stone, "Top 10 US Retailers: Amazon Joins Ranks of Walmart, Kroger for First Time Ever," *International Business Times,* July 3, 2014, http://www.ibtimes.com/top–10-us-retailers -amazon-joins-ranks-walmart-kroger-first-time-ever–1618774.

15. http://www.businessinsider.com/why-wal-marts-pay-is-lower-than -costco-2014-10.

16. Sinegal, interview, 2015. See also, Megan McArdle, "Why Wal-Mart Will Never Pay Like Costco," *Bloomberg View,* August 27, 2013, http:// www.bloombergview.com/articles/2013-08-27/why-walmart-will-never -pay-like-costco.

17. Aaron Taube, "Why Costco Pays Its Retail Employees $20 an Hour," *Business Insider,* October 23, 2014, http://www.businessinsider.com /costco-pays-retail-employees–20-an-hour–2014–10; Mitch Edelman, "Wal-Mart Could Learn from Ford, Costco," *Carroll County Times,* July 19, 2013, http://www.carrollcountytimes.com/cct-arc-67d6db6e-db9f -5bc4-83c3-c51ac7a66792-20130719-story.html.

18. Wayne F. Cascio, "The High Cost of Low Wages," *Harvard Business Review,* December 2006 issue, https://hbr.org/2006/12/the-high-cost-of -low-wages; for more information on this strategy, see Zeynep Ton, "Why 'Good Jobs' Are Good for Retailers," *Harvard Business Review,* January– February 2012, https://hbr.org/2012/01/why-good-jobs-are-good-for -retailers/?conversationId=3301855.

19. Sinegal, interview, 2015.

20. Sinegal, interview, 2015.

21. Saritha Rai, "The Fifth Metro: Doing IT Differently," *The Indian Express,* November 24, 2014, http://indianexpress.com/article/opinion/columns /the-fifth-metro-doing-it-differently/.

22. *Zoho,* https://www.zoho.com/; see also "Sridhar Vembu," *Wikipedia,* April 17, 2015, https://en.wikipedia.org/wiki/Sridhar_Vembu.

23. *Zoho,* https://www.zoho.com/.

24. Mark Milian, "No VC: Zoho CEO 'Couldn't Care Less for Wall Street'," *Bloomberg,* November 29, 2012, http://go.bloomberg.com /tech-deals/2012-11-29-no-vc-zoho-ceo-couldnt-care-less-for-wall-street/; Rasheeda Bhagat, "A Life Worth Living," *Rotary News,* October 1, 2014, http://www.rotarynewsonline.org/articles/alifeworthliving.

25. Sridhar Vembu, interviewed by Todd Rose, April 21, 2015; see also: Rasheeda Bhagat, "Decoding Zoho's Success," *The Hindu Business Line,* February 4, 2013, http://www.thehindubusinessline.com/opinion /columns/rasheeda-bhagat/decoding-zohos-success/article4379158.ece.

26. Vembu, interview, 2015.

27. Vembu, interview, 2015.

28. Vembu, interview, 2015; for similar sentiments, see Sridar, "How We Recruit—On Formal Credentials vs. Experience-based Education," *Zoho Blogs,* June 12, 2008, http://blogs.zoho.com/2008/06/page/2.

29. *Zoho University,* http://www.zohouniversity.com/; Bhagat, "A Life Worth Living."

30. Vembu, interview, 2015.

31. Vembu, interview, 2015.

32. Vembu, interview, 2015. See also "Zoho University Celebrates a Decade of Success," https://www.zoho.com/news/zoho-university-celebrates -decade-success.html; Leslie D'Monte, "Challenging Conventional Wisdom with Zoho University," *Live Mint,* November 21, 2014, http://www .livemint.com/Companies/LU4qIlz47C5Uph2P5i250K/Challenging -conventional-wisdom-with-Zoho-University.html.

33. Krithika Krishnamurthy, "Zoho-Run Varsity Among Its Largest Workforce Providers," *Economic Times,* March 14, 2014, http://articles .economictimes.indiatimes.com/2015–03–14/news/60111683_1 _students-csir-iisc.

34. Vembu, interview, 2015; D'Monte, "Challenging Conventional Wisdom."

35. Vembu, interview, 2015.

36. Vembu, interview, 2015.

37. Vembu, interview, 2015.

38. Vembu, interview, 2015.

39. Vembu, interview, 2015.

40. "About Us: Company History," *The Morning Star Company,* http://morning

starco.com/index.cgi?Page=About%20Us/Company%20History.

41. See "About Us: Company History"; Frédéric Laloux, *Reinventing Orga-nizations: A Guide to Creating Organizations Inspired by the Next Stage of Human Consciousness* (Brussels: Nelson Parker, 2014), 112; and "Chris Rufer," http://www.self-managementinstitute.org/about/people/1435.

42. See Allen, "Passion for Tomatoes," "About Us: Company History."

43. Laloux, *Reinventing Organizations,* 112; Goldsmith, "Morning Star Has No Management."

44. Paul Green Jr., interviewed by Todd Rose, July 28, 2014.

45. "About Us: Colleague Principles," *The Morning Star Company,* http://morningstarco.com/index.cgi?Page=About%20Us/Colleague%20 Principles.

46. Gary Hamel, "First, Let's Fire All the Managers," *Harvard Business Review,* December 2011, https://hbr.org/2011/12/first-lets-fire-all-the -managers.

47. Green, interview, 2014.

48. Green, interview, 2014.

49. Green, interview, 2014.

50. Green, interview, 2014.

51. Green, interview, 2014.

52. Green, interview, 2014.

53. Sinegal, interview, 2015.

54. Vembu, interview, 2015.

CHAPTER 8: REPLACING THE AVERAGE IN HIGHER EDUCATION

1. For an overview of the problems and the opportunities, see Michelle R. Weise and Clayton M. Christensen, *Hire Education: Mastery, Modular-ization, and the Workforce Revolution* (Clayton Christensen Institute, 2014), http://www.christenseninstitute.org/wp-content/uploads/2014/07 /Hire-Education.pdf.

2. Casey Phillips, "A Matter of Degree: Many College Grads Never Work in Their Major," *TimesFreePress.com,* November 16, 2014, http://www .timesfreepress.com/news/life/entertainment/story/2014/nov/16/matter -degree-many-college-grads-never-work-/273665/.

3. James Bessen, "Employers Aren't Just Whining—The 'Skills Gap' Is Real," *Harvard Business Review*, August 25, 2014, https://hbr .org/2014/08/employers-arent-just-whining-the-skills-gap-is-real; Stephen Moore, "Why Is It So Hard for Employers to Fill These Jobs?" CNSNews .com, August 25, 2014, http://cnsnews.com/commentary/stephen-moore /why-it-so-hard-employers-fill-these-jobs.

4. Jeffrey J. Selingo, "Why Are So Many College Students Failing to Gain Job Skills Before Graduation?" *Washington Post,* January 26, 2015, www .washingtonpost.com/news/grade-point/wp/2015/01/26/why-are-so -many-college-students-failing-to-gain-job-skills-before-graduation/; Eduardo Porter, "Stubborn Skills Gap in America's Work Force," *New York Times,* October 8, 2013, http://www.nytimes.com/2013/10/09 /business/economy/stubborn-skills-gap-in-americas-work-force .html; and Catherine Rampell, "An Odd Shift in an Unemployment Curve," *New York Times,* May 7, 2013, http://economix.blogs.nytimes .com/2013/03/07/an-odd-shift-in-an-unemployment-curve/.

5. Michelle Jamrisko and Ilan Kolet, "College Costs Surge 500% in U.S. Since 1985: Chart of the Day," *Bloomberg Business,* August 26, 2013, http://www.bloomberg.com/news/articles/2013–08–26/college-costs -surge–500-in-u-s-since–1985-chart-of-the-day.

6. Jamrisko and Kolet, "College Costs Surge 500% in U.S. Since 1985."

7. "Making College Cost Less," *The Economist,* April 5, 2014, http://www .economist.com/news/leaders/21600120-many-american-universities -offer-lousy-value-money-government-can-help-change; "Understand- ing the Rising Costs of Higher Education," *Best Value Schools,* http:// www.bestvalueschools.com/understanding-the-rising-costs-of-higher -education/.

8. Raymond E. Callahan, *Education and the Cult of Efficiency* (Chicago: University of Chicago Press, 1964).

9. Judy Muir, interviewed by Todd Rose, October 28, 2014. For more infor- mation about Muir's approach to college admissions, see Judith Muir and Katrin Lau, *Finding Your U: Navigating the College Admissions Process* (Houston: Bright Sky Press, 2015).

10. Muir, interview, 2014.

11. Bill Fitzsimmons, interviewed by Todd Rose, August 4, 2014.

12. Elena Silva, "The Carnegie Unit—Revisited," *Carnegie Foundation,* May 28, 2013, http://www.carnegiefoundation.org/blog/the-carnegie-unit -revisited/.

13. For a broader critique of diplomas, see Charles A. Murray, "Reforms for

the New Upper Class," *New York Times,* March 7, 2012, http://www
.nytimes.com/2012/03/08/opinion/reforms-for-the-new-upper-class.html.

14. "Micro-Credentialing," *Educause,* http://www.educause.edu/library
/micro-credentialing; and Laura Vanderkam, "Micro-credentials," *Laura
Vanderkam,* December 12, 2012, http://lauravanderkam.com/2012/12
/micro-credentials/.

15. Gabriel Kahn, "The iTunes of Higher Education," *Slate,* September
19, 2013, http://www.slate.com/articles/technology/education/2013/09
/edx_mit_and_online_certificates_how_non_degree_certificates
_are_disrupting.html; https://www.edx.org/press/mitx-introduces
-xseries-course-sequence; Nick Anderson, "Online College Courses to
Grant Credentials, for a Fee," *Washington Post,* January 9, 2013, http://
www.washingtonpost.com/local/education/online-college-courses
-to-grant-credentials-for-a-fee/2013/01/08/ffc0f5ce–5910–11e2
–88d0-c4cf65c3ad15_story.html; Nick Anderson, "MOOCS—Here
Come the Credentials," *Washington Post,* January 9, 2013, http://www
.washingtonpost.com/blogs/college-inc/post/moocs—here-come-the
-credentials/2013/01/09/a1db85a2–5a67–11e2–88d0-c4cf65c3ad15
_blog.html.

16. Maurice A. Jones, "Credentials, Not Diplomas, Are What Count for
Many Job Openings," *New York Times,* March 19, 2015, http://www
.nytimes.com/roomfordebate/2015/03/19/who-should-pay-for-workers
-training/credentials-not-diplomas-are-what-count-for-many-job-openings;
for more on national credential initiative, see "President Obama and Skills
for America's Future Partners Announce Initiatives Critical to Improving
Manufacturing Workforce," *The White House,* June 8, 2011, https://www
.whitehouse.gov/the-press-office/2011/06/08/president-obama-and-skills
-americas-future-partners-announce-initiatives.

17. Jones, "Credentials, Not Diplomas."

18. http://www.slate.com/articles/technology/education/2013/09/edx_mit
_and_online_certificates_how_non_degree_certificates_are_disrupting
.html.

19. Thomas R. Guskey, "Five Obstacles to Grading Reform," *Educational
Leadership* 69, no. 3 (2011): 16–21.

20. Western Governors University, http://www.wgu.edu/.

21. "Competency-Based Approach," Western Governors University,
http://www.wgu.edu/why_WGU/competency_based_approach?utm
_source=10951; John Gravois, "The College For-Profits Should Fear,"
Washington Monthly, September/October 2011, http://www.washington

monthly.com/magazine/septemberoctober_2011/features/the_college
_forprofits_should031640.php?page=all; "WGU Named 'Best Value
School' by University Research & Review for Second Consecutive Year,"
PR Newswire, April 9, 2015, http://www.prnewswire.com/news-releases
/wgu-named-best-value-school-by-university-research—review-for
-second-consecutive-year–300063690.html; Tara Garcia Mathewson,
"Western Governors University Takes Hold in Online Ed," *Education
DIVE,* March 31, 2015, http://www.educationdive.com/news/western
-governors-university-takes-hold-in-online-ed/381283/.

22. George Lorenzo, "Western Governors University: How Competency-
Based Distance Education Has Come of Age," *Educational Pathways* 6,
no. 7 (2007): 1–4, http://www.wgu.edu/about_WGU/ed_pathways_707
_article.pdf; Matt Krupnick, "As a Whole New Kind of College Emerges,
Critics Fret Over Standards," *Hechinger Report,* February 24, 2015,
http://hechingerreport.org/whole-new-kind-college-emerges-critics-fret
-standards/.

23. Krupnick, "As a Whole New Kind of College Emerges;" and "Overview," Competency-Based Education Network, http://www.cbenetwork
.org/about/.

24. EdX and Arizona State University Reimagine First Year of College, Offer
Alternative Entry Into Higher Education," April 22, 2015, https://www
.edx.org/press/edx-arizona-state-university-reimagine; John A. Byrne,
"Arizona State, edX to offer entire freshman year of college online," *Fortune,* April 22, 2015, http://fortune.com/2015/04/22/arizona-state-edx
-moocs-online-education/. For more on ASU, see Jon Marcus, "Is Arizona
State University the Model for the New American University?" *Hechinger Report,* March 11, 2015, http://hechingerreport.org/is-arizona-state
-university-the-model-for-the-new-american-university/.

CHAPTER 9: REDEFINING OPPORTUNITY

1. For more information about the A-10 Warthog, see "Fairchild Republic
A-10 Thunderbolt II," *Wikipedia,* June 29, 2015, https://en.wikipedia.org
/wiki/Fairchild_Republic_A–10_Thunderbolt_II.

2. Lt. Kim C. Campbell, interviewed by Todd Rose, April 8, 2015.

3. Campbell, interview, 2015.

4. Campbell, interview, 2015.

5. Campbell, interview, 2015.

6. Campbell, interview, 2015.

7. "Kim Campbell," *Badass of the Week,* April 7, 2003, http://www.badass oftheweek.com/kimcampbell.html.

8. "Kim N. Campbell," *Military Times,* http://valor.militarytimes.com /recipient.php?recipientid=42653.

9. Campbell, interview, 2015.

10. Campbell, interview, 2015.

11. Campbell, interview, 2015.

12. For an overview of the concept of equal opportunity, see "Equal Opportunity," *Wikipedia,* June 24, 2015, https://en.wikipedia.org/wiki/Equal _opportunity.

13. Equal access has played a profoundly important role in the fight for equality based on race (see "School Desegregation and Equal Education Opportunity," *Civil Rights 101,* http://www.civilrights.org/resources /civilrights101/desegregation.html?referrer=https://www.google.com/, and "The Civil Rights Movement (1954–1965): An Overview," *The Social Welfare History Project,* http://www.socialwelfarehistory.com/eras /civil-rights-movement/); gender (see Bonnie Eisenberg and Mary Ruthsdotter, "History of the Women's Rights Movement," *National Women's History Project,* 1998, http://www.nwhp.org/resources/womens-rights -movement/history-of-the-womens-rights-movement/); and disability ("A Brief History of the Disability Rights Movement," *The Anti-Defamation League,* 2005, http://archive.adl.org/education/curriculum_connections /fall_2005/fall_2005_lesson5_history.html.)

14. It is crucial here to recognize that equal access still matters and is worth fighting for. Take, for example, the fact that in 2005 (two years after Killer Chick's heroics), there was an effort in Congress to bar women from combat ("Letters to the Editor for Friday, May 27, 2005," *Stars and Stripes,* May 27, 2005, http://www.stripes.com/opinion/letters -to-the-editor-for-friday-may-27-2005-1.35029).

15. Abraham Lincoln, "Message to Congress," July 4, 1861, *Collected Works of Abraham Lincoln,* vol. 4 (Rutgers University Press, 1953, 1990): 438.

16. For more information on norm-referenced tests, see "Norm-Referenced Achievement Tests," *FairTest,* August 17, 2007, http://www.fairtest.org /norm-referenced-achievement-tests.

17. James Truslow Adams, *The Epic of America* (New York: Blue Ribbon, 1931), 214–215.

18. Adams, "Epic of America," 180.

INDEX

of failure of single-score, 84–85; weak correlations problem of, 85–91; Zoho Corporation's avoidance of, 157. *See also* average measurement; student ranking
Religious Education Association, 110
remedial math class, 140, 141
Renault (France), 48
Rockefeller, John D., 50
Roosevelt, Franklin D., 48
Rufer, Chris, 158

Salesforce.com, 153, 154
Sam's Club, 153
SAT scores: college admissions process and role of, 168; Google hiring decisions consideration of GPAs and, 77–78, 79, 85; weak correlation problem in performance and, 90–91
schools: Bloom's research findings on benefits of individual pace to learning in, 131–33; "Gary Plan" adopted by U.S., 51; General Education Board's essay (1912) supporting Taylorism, 50–51; history of the educational Taylorism of American, 50–52; knowledge on if-then signatures used to change approach to, 108–9; the pros and cons of the Taylorization of our, 56–58; as Queteletian process turning Average Students into Average Workers, 52; studies challenging the assumption of speed and learning used by, 131–35; success of Khan Academy's self-paced modules for student learning, 133–34; Thorndike's assumption that faster equals smarter and development of standardized, 130–31; Thorndike's use of rank concept to separate superior and inferior students, 52–56; Zoho Corporation's Zoho University acceptance of impoverished Indian students, 155–56, 158. *See also* American educational system; higher education
Science Careers magazine, 139
science of the individual: challenge to overcoming averagarianism to *analyze, then aggregate* of, 69–72; description of the, 12; early concerns over the rise of averagarians and average measurements in, 36–38; Galton's idea of "law of deviation from the average" and concept of rank, 32–35, 36; on individual pace and sequences of individual progress, 129–43; Molenaar's *aha-erlebnis* ("epiphany") on the ergodic switch implications for, 65–69; Quetelet's Average Man studies and concept of type, 26–31, 36. *See also* individuality
scientific career of excellence pathway, 138–39
scientific management ("Taylorism"): applied to the American educational system, 49–52, 124–25, 167; historic development of, 42–45; human resources born out of, 117–18; James Truslow Adams' criticism of, 189–90; Morning Star Company's rejection of, 159; *The Principles of Scientific Management* (Taylor) on, 47; reinforcing normative thinking, 124–25; superiority of the principles of individuality's win-win capitalism over, 163–64; widespread and global adoption of the, 47–49; worker apathy problem as legacy of, 147–49. *See also* averagarianism; Taylor, Frederick Winslow
Scottish soldiers "average" study (1840s): description of Quetelet's, 26–28; Molenaar's epiphany on Quetelet's misinterpretation of his, 64–65
Second Industrial Revolution, 41
self-control: Kidd's study adding context to the marshmallow study on, 113–14; Mischel's marshmallow study on, 112; Shoda and Mischel's follow-up to marshmallow study on, 112–13
self-determined educational pathways: alignment of credentialing and, 178; common reactions when first introduced to idea of, 178–79; competency-based evaluations alignment with, 178; potential benefits of adopting a, 179–82; transformation of higher education through, 170, 176–79
self-paced learning: as benefiting all students, 134–35; Bloom's research on student learning benefits of, 131–33; Khan Academy's successful modules allowing, 133–34; pathways principle affirmation on the, 129–35. *See also* developmental sequences

ABOUT THE AUTHOR

Todd Rose is the director of the Mind, Brain, and Education program at the Harvard Graduate School of Education, where he also leads the Laboratory for the Science of the Individual. He is also the cofounder and president of The Center for Individual Opportunity, a nonprofit organization that promotes the principles of individuality in work, school, and society. Todd lives in Cambridge, Massachusetts.